WARGASM

The Psychology of Warfighting and Warfighters

By Nicholas L. Long

WARGASM

The Psychology of Warfighting and Warfighters
© 2025 Nicholas Long
Published by Fortes Via Press

Contact Information nicholaslongwritings@gmail.com

All Rights Reserved

Printed in the United States

Second Edition

No part of this publication may be reproduced, distributed, or transmitted in any form or by any means, including photocopying, recording, or other electronic or mechanical methods, now known or to be known in the future, without the prior written permission of both the publisher and the author, except in the case of brief quotations embodied in critical reviews and certain other noncommercial uses permitted by copyright law.

ISBN PRINT - 979-8-9994929-3-7

Wargasm: the release of built-up tension from the continuous stress of combat, resulting in a sense of clarity and control.

Dedication

This book is dedicated to the memories of the Marines that we lost in Afghanistan and Iraq:

                LCpl Brian A. Medina
                LCpl Brian C. Hopper
                LCpl William J. Leusink
                Cpl Nicholas C. Kirven
                Cpl Richard P. Schoener
                Sgt David R. Christoff
                Sgt Joseph M. Garrison
                Sgt Ryan H. Lane
                SSgt Jason C. Ramseyer

Semper Fidelis Brothers

Table of Contents

Dedication 4

Prologue: The Eternal Shadow of War 8

Chapter 1: War Is My Refuge 12

Chapter 2: The Chemistry of Combat 22

Chapter 3: The Fog of War 30

Chapter 4: The Strength of the Wolf Is the Pack 46

Chapter 5: The Tribe We Leave Behind 60

Chapter 6: Calm in the Chaos 72

Chapter 7: Lessons from the Battlefield 82

Chapter 8: Embrace the Suck 92

Chapter 9: Reclaiming the Warrior Spirit 100

References 114

About the Author 130

Prologue: The Eternal Shadow of War

The existence of war throughout human history extends beyond military confrontations and strategic maneuvers, serving as a transformative test for human resilience and spirit. Each historical period has seen conflict compel us to examine our core identity with the questions: Who are we? What drives us? What truths about our human essence can moments of chaos demonstrate?

The persistent attraction to war exerts a powerful influence that molds both historical events and the personal journeys of warriors who enter combat zones. The war experience presents warfighters with contrasting realities. Devastation brings clarity, while chaos reveals purpose. The fundamental survival instincts from ancient times have evolved to become a deep source of guidance when the world presents insurmountable uncertainty.

This book presents war objectively without glorifying its violence or understating its devastating impact. It explores the psychological processes that bridge instinctual behavior with learned experience. This analysis reveals how the same forces that lead us to engage in conflict also provide us with perseverance, strength, and leadership focus while turning hardship into growth opportunities. Readers who have experienced combat or studied it from a distance will find these pages urging them to unearth the deeper connections between violence and the complex dynamics of survival and purpose.

During this journey, you will learn that combat clarity stems from innate instincts developed over time to sharpen our focus when facing severe stress. When properly nurtured and redirected, these instincts become powerful tools for personal transformation and building a meaningful life after military service.

The upcoming narrative combines elements of personal reflection with historical context alongside insights from military and civilian viewpoints. During my combat service, I experienced terror and uncertainty, but through these moments, I achieved deep connections to my higher purpose.

After experiencing battlefield clarity, I faced reintegration difficulties in civilian life, which were unexpectedly resolved by my subsequent return to combat. I weave these stories into the pattern of a larger tapestry that reveals how conflict shapes the human spirit.

This prologue asks readers to explore war as both an external confrontation and an internal struggle against our fundamental fears and drives while also serving as a chance for personal renewal. The impact of war reaches beyond military history by shaping both our collective identity and the very beat of our hearts.

The upcoming exploration will demonstrate how warfighting psychology affects warriors' mental states and will show how these basic instincts can be consciously transformed into beneficial activities. It is a guide for those who seek to understand the often-contradictory nature of conflict: Conflict holds the power to destroy and to empower human beings while simultaneously inflicting wounds and providing healing before drawing us toward destruction yet ultimately guiding us toward creation and renewal.

In unraveling these themes, we confront our most enduring question: What methods enable us to transform conflict's intense energy into building both a sustainable existence and enduring peaceful progress? The solutions to this dilemma are complex and diverse, yet each revelation forms a building block toward understanding our instinctual dualities.

I hope you approach this exploration of battlefield psychology prepared to discover how chaos yields both remarkable mental clarity and serious sacrifices. The text stands as a tribute to the life-changing potential found in facing and accepting our most basic instincts instead of glorifying violent behavior. The constant presence of conflict can teach us how to illuminate our way toward living with more meaning.

Chapter 1: War Is My Refuge

Ask any combat veteran from any war or battle throughout history, and they will likely say the same thing: No experience delivers a higher adrenaline rush than the intensity of a firefight. While there are many out there who will say fighting is intrinsically evil, the reality is that war and violence are primal instincts of survival that endure as the only reality under which you defend yourself or loved ones. There is this imperative hardwired in our DNA. To Sigmund Freud, violence is a life-preserving instinct, not simply a destructive force. The psychologist E.O. Wilson called warfare "humanity's hereditary curse" and said it transcends any culture or any period because it is "universal and eternal."

The origins of violence and war go back to the first killing, when Cain killed his brother Abel. Since that moment, conflict has become part of the fabric of humanity, weaving its way into the narratives of how we build society and identity. Military historian John Keegan expressed this duality beautifully when he said,

> *"Warfare is almost as old as man himself, and reaches into the places of the human heart, places where self dissolves rational purpose, where pride reigns, where emotion is paramount, where instinct is king."*

Even religious texts that tell us the importance of love and peace tell us war and violence are part of who we are. In the Bible, James 4:1 tells us that war and fighting come from something inside of us. Similarly, Matthew 5:9 says, "Blessed are the peacemakers, for they will be called children of God." And indeed, ironically, if you ask any warfighter, they think of themselves as peacemakers, maintaining peace through planned acts of violence. Conflict is not just eternal, as Wilson suggests—it is internal, inherently so. After all, world history has been punctuated by the opening and closing of conflicts, and much of humanity's great technological invention has taken place due to military necessity.

Nobody in their right mind would condone random violence against their fellow humans, but there appears to be a fascination, if not obsession, with violence and war. Our pop culture is littered with it. The most popular books, movies, music, and video games are often built around monumental themes like violence, combat, and survival. The lure of such phenomena is telling, according to psychologist and author Lawrence LeShan:

> "War sharpens experience, heightens perceptions, and makes one more and more aware of one's own existence. At the same time, war offers us the opportunity to become part of something larger and more intense-to be a note that has blended into a symphony."

Many psychological researchers have observed thrill seeking activities among soldiers returning from war, often called "the adrenaline junkie phenomenon." This phenomenon typically manifests itself in the form of reckless driving, extreme sports, or substance abuse. Often, warfighters look to return to war like a drug addict in search of the next high. From a physiological perspective, combat releases heightened levels of dopamine in the brain and increased testosterone levels throughout the body. The emotions, thoughts, and instincts are raw and basic, making them highly addictive. There is something pure about combat. Knowing the other side wants to kill you, and you them, nothing else matters. All that matters is survival for oneself and their tribe. As chaos swirls around them, warfighters maintain calm. Social influences get stripped away, allowing emotions, thoughts, and instincts to retain their natural intensity. In this sense, war is life stripped to its most raw, essential elements.

But the toughest fight sometimes comes after the warfighting life ends. Returning to civilian life finds us jolted into matters of the mundane. Things like everyday bills, groceries, child-rearing, 401(k)s, and gas prices become the complex worries and concerns of everyday life. What used to seem chaotic now seems comforting in its simplicity by comparison. Conversely, there are some who think all of those everyday worries are meaningless when compared to what our warfighters go through. But to many soldiers, the high of combat is much more appealing than the monotony of normalcy. After hanging up their boots, many warriors are haunted by the memory of a life lived in extremes, never knowing if they really lived that life or if it was all a dream. They seek out new ways to fill that sensation. Those who are unable to find it continue to seek new avenues, often to their own detriment. When I left the Marines, I fell into a dark pattern of drinking. Relationships failed. I was a ship without a rudder. After almost two years, when the Marine Corps came calling to say I was needed again, it was almost a relief. I would be returning to the only world I'd known since I was just an 18-year-old kid. War was my refuge, bringing purpose and meaning back to life.

From personal experience, I can tell you that the battles I fought, the horrors I've seen, replay in my head as if it was a movie I'd watched, rather than an experience I actually lived. This sense of detachment causes many problems and makes the reassimilation process much more difficult. Are the memories real? As the decades go on, it seems even more distant from reality. More than once I have had conversations with Marines with whom I served, and the details of battles fought don't add up. Memories are malleable and prone to distortion. Research has found that traumatic memories are often repeatedly rehearsed, leaving the mind susceptible to altering the memories. This can make them seem even less real.

This paradox of war, tormenting one instant and then providing refuge the next, isn't new or confined to the War on Terror. It's an inheritance of millions of generations, inscribed in the bedrock of our past, our mythology, our identity. To see why our human ancestors kept returning to war, we need to step outside the avenues of contemporary warfighters and apparitions of battlefields and trenches of modernity. We need to turn again to the myths that are stitched into the deep fiber of our present and our future.

The theme of conflict is so universal to humanity that it perhaps should be an assumption. It weaves through our earliest myths, our oldest historical records, and our collective psyche. Through sermons and psalms, blown trumpets, and boiled texts, we tell and retell the same old story of the same old war, the same old enemy, the same old rivalry. We do it over and over again, as if they were already a perennial and unalterable problem of human beings, generation after generation. We may discover something worthwhile in studying these mythic narratives about our stubborn freight toward conflict and violence.

Mythology: What Do We Learn from Conflict? Archetypes

As I've just described, one of the earliest stories of conflict we have recorded in any format is the story of Cain and Abel, and it is a narrative as primal and universal as they come. The dynamic between two brothers extends beyond time and culture. Cain's jealousy, which was intensified by his own self-worthlessness and sense of rejection, ends up causing the death of Abel. This story gives us an archetypal theme — conflict as reflexive, a manifestation of our inward struggles, with feelings like fear, envy, and pride marching down the street.

This theme is echoed across many other cultures as well. In Roman mythology, the brothers Romulus and Remus were destined for greatness, but their competition came to a bloody end, with Romulus killing his brother and (according to legend), founding Rome. The story is a violent parable of the origins of power, order, and identity.

The comparison can also be drawn from the Indian epic Mahabharata, which deals with two feuding families, the Pandavas and the Kauravas, grappling with concepts of righteousness, duty, and justice. These stories not only reveal insights about the nature of their respective societies but also reflect humanity's psychological and emotional struggles.

Myths from across time and space provide us with a common ground for understanding the inevitability, power, and price of conflict. The Norse sagas tell of gods against gods, such as

Loki and the Aesir, chaos versus order. In these tales, conflict is not merely slaying, it is also metamorphosis, a means of reconstructing the self and society.

War in Early Human Societies

Stories beyond mythology emerge from the remains of prehistoric wars. Archaeological finds, such as mass grave sites, ancient fortifications, and crude weapons, show that early human tribes fought for resources, survival, and territory. These wars did not include the massed combats of later cultures but were conflicts of necessity and instinct. They reveal how, at base, conflict became a tool to ensure survival and define the boundaries of groups. They show us that conflict, violence, and even war, are elements of the natural state of humanity.

Territories aren't the only thing that humans fight over, and territorial disputes aren't limited to people, as other species clash for similar reasons. But what distinguishes humans is that we ritualize and immortalize these fights, committing them to story so they serve as lessons for others who come after us. These early conflicts helped lay the foundation for what would become social orders, with leadership, loyalty, and strategy forming the basics of society. Every culture in history has its tales of glory and battle. Widespread songs and sagas were used to keep alive the memory of them, even when the very cultures eventually faded away.

Symbolism and Ritual: The Ceremonial Aspect of Conflict

For ancient people, war had physical, symbolic, and ceremonial elements. Conflict was often linked to religious and cultural rituals that gave it meaning. War played a central role in Aztec culture, deeply embedded within their worldview and a critical aspect of their spiritual and cosmic beliefs. Therefore, human sacrifices in wars were thought to be offerings to the gods for the continuation of the universe.

Much like the Maori and Zulu had their war dances, which served not just to show force, but also to condition the psyche before entering a battle, creating a collective identity and purpose for warriors. These rituals show how humanity has long tried to straddle the chaos of conflict with the order of culture and meaning. They provided a framework for giving fear purpose, lent resonance to acts of courage, cast violence in the context of a larger mission and their own survival.

The Evolution of Conflict Narratives

As societies developed, their stories of conflict evolved, too. Historical texts, such as Homer's Iliad and the Indian Mahabharata, have a mythical lens, interspersing deities and heroes who have a foot in both worlds with their human encounters. Therefore, these epics teach us that war is not merely about battle sequences, but a moral and existential struggle. They provide lessons on honor, sacrifice, and the human price of war that echo just as loudly today as they did millennia ago.

Modern narratives, from Shakespeare's histories to contemporary war films, continue to draw from these ancient themes. They resonate because the instinct for conflict and the search for meaning within it remain universal. Whether through the tragedies of Cain and Abel or the victories of Romulus, these legends remind us that the battles we fight are as old as humanity itself, and as much a part of the human existence as any other element of life.

While the firelight of combat illuminates the raw, primal nature of humanity, its embers burn deep in the stories and histories that have defined us. The instincts that drive warriors to find refuge in war are not isolated to the individual, but instead are part of a broader, shared legacy that stretches back through the beginning of time. Whether etched in myth or recorded in history, our ancestors reveal how warfare and survival are inextricably linked to the core of our being. The escape that war offers is not merely individual experience, but part of a larger story that is written indelibly into the story of being human. The primal instincts that drive warfighters—the adrenaline, the focus, the raw simplicity of survival—are echoes of a shared inheritance, shaped by centuries of conflict and reflected in the myths and histories that define us. These are base-level instincts forged and legitimized in tidal waves of centuries-old blood, reiterated in the myths and histories that animate us.

However, mythology is only one side of the coin. As is so often the case, the historical record shows that the instincts sharpened in the chaos of battle served a wider purpose to nascent human societies. They built loyalty, formed communities, and accelerated technology, even as they colored the psyche of people who fought. This is how the individual and humanity tie together. The instincts that propel a warrior to war are also the forces that have impelled civilizations to defend, grow, and persist. War is simultaneously intensely personal and massively universal and is a question of chaos and purpose that connects the inner fight of the single soul to the large course of human history.

What We Can Take from This

War is not only a feature of the outside world. It is more than is credited through the famous quote by Prussian General Carl von Clausewitz, "war is a continuation of policy by other means." Its roots run deep as an integral part of human nature, embedded in the predatory drives that have defined humanity throughout history. For the modern warfighter, this ancient drive elucidates the deep-seated peace, calm, and purpose they discover amid the tumult of battle. In the unadulterated black and white of survival and the tribal tether of war, they rise to a greatness that dances on the threshold of time and space. However, this very sensation leaves a longing for that heightened focus and directedness long after the dust of the battlefield has settled, illustrating the paradox of war as both a place to take refuge and a source of intellectual tumult.

Chapter 2: The Chemistry of Combat

The human body undergoes remarkable changes when engaged in battle. The body enlists every nerve and muscle fiber alongside each cell to endure as primal instincts forcefully dominate our actions. The body's physiological response to combat goes beyond basic danger reflexes because it produces an intensified state of alertness and concentration which warfighters often find desirable. The physiological changes that enhance battlefield performance can become addictive because soldiers develop a desire to experience combat's unique state of focused clarity. This chapter examines how physiological changes in the warfighter's body create an attraction to conflict which many find hard to escape.

The Onset of the Fight-or-Flight Response

The hypothalamus in the brain immediately transmits an urgent signal via the autonomic nervous system, which activates the adrenal glands to produce adrenaline (epinephrine) and cortisol when a threat is perceived. Cortisol works to regulate energy levels and crucially supports energy supply during aggressive or combat situations. Neurons and muscles primarily derive their energy from glucose. Blood sugar levels rise because cortisol stimulates glucose production and ensures sufficient energy supplies for extended defensive or evasive actions in stressful conditions. These hormones trigger the familiar fight-or-flight response within seconds. Adrenaline rushes through the bloodstream as it increases heart rate, along with blood pressure and breathing rate, while enhancing vision and mental focus.

The quick mobilization of the body achieves optimal performance rather than simply demonstrating brute strength. During these pivotal moments every bodily function becomes dedicated to one goal: survival. The physiological surge contains a paradox because warfighters find this intense and clear state both exciting and addictive. The heightened clarity and focus people find during extreme stress show them a purer way of living, which many people continue to chase after the battle ends. The organism's endurance is extended during active alert periods through energy conservation that supports both immune and integrative bodily functions. Elevated cortisol levels produce short-term benefits, yet prolonged exposure to increased cortisol may produce adverse health effects, which underscores the need for effective stress management techniques during combat situations.

Cardiovascular and Respiratory Transformations

During combat situations, the cardiovascular system experiences significant changes that support optimal physical performance. The human body enhances the circulation of oxygenated blood to critical organs and muscles through adrenaline-triggered heart rate acceleration. The cardiovascular system acts instantly when adrenaline flows through the body. The respiratory system speeds up simultaneously to handle the increased need for oxygen. The diaphragm works very hard to enable maximum oxygen intake during rapid and deep breaths. Combat readiness depends on essential modifications to the cardiovascular system. The sudden increase in blood flow speeds up circulation and directs blood to essential muscle groups to keep the body ready for quick and powerful movements.

Peak performance demands a crucial interaction between cardiac and pulmonary systems. Muscle groups powering the warfighter receive necessary fuel for intense operations while the brain maintains alertness and readiness for decision-making. The short-term benefits of these adaptations prove vital for warfighters who eventually discover that maintaining high-performance levels permanently affects them. The fleeting sensation of peak system performance becomes addictive, making many yearn to regain that exceptional mental clarity and physical strength.

Prolonged combat stress heavily burdens the cardiovascular system, which may result in harmful health effects. The constant flow of adrenaline and increased heart rate during extended periods of heightened alertness puts persistent strain on the heart, which raises the chance of cardiovascular disease development. The continuous stress response puts blood vessel pressure under constant strain to keep the body ready, which contributes to the development of hypertension and accelerates cardiovascular deterioration over time. The constant presence of stress hormones like cortisol can interfere with normal cardiovascular operations and worsen health problems for individuals in combat situations and beyond. These long-term impacts emphasize the critical need for proper stress management measures to protect soldiers from cardiovascular risks in high-stress situations.

Metabolic Shifts: Fueling the Fire

The body's energy production and metabolic activities become the main focus during combat situations. When engaged in battle, our bodies undergo major metabolic changes, primarily through the rise of cortisol levels in our bloodstream. When cortisol affects the body, the liver activates to break down glycogen reserves and produces glucose, resulting in an instant surge of energy. Through this process, both brain and muscle maintain optimal performance levels throughout extended active periods.

Cortisol maintains energy levels allowing warfighters to demonstrate exceptional endurance after adrenaline provides an initial energy surge. The physiological cost matches that of the emotional highs experienced during combat. The extended presence of high cortisol levels destroys typical metabolic functions which may result in persistent health issues including insulin resistance and type 2 diabetes. Extended periods of cortisol secretion result in challenges including fatigue, metabolic disturbances, and unstable weight patterns. The powerful feeling that comes from being fueled by a continuous flow of energy becomes synonymous with mental clarity for many people who experience it as proof of testing limits and surviving the process. The implementation of successful stress management and metabolic health maintenance strategies is vital for reducing these potential risks and protecting against the long-term negative effects from combat-induced physiological changes.

Sensory Enhancements and the Distillation of Perception

Sensory enhancements serve as a crucial evolutionary benefit in combat scenarios by equipping individuals to identify and react to threats with improved detection ability. The surge of adrenaline and stress hormones leads to substantial changes in both visual and auditory systems by enhancing sight clarity and boosting hearing sensitivity. The advanced visual skills soldiers develop allow them to notice slight movements and environmental changes which are vital for enduring rapidly changing combat situations. Enhanced hearing capabilities allow people to perceive quieter sounds which aids them in detecting and locating potential threats despite noisy surroundings. The immediate tactical benefits of these enhancements come at the cost of potential long-term negative effects like sensory fatigue and hearing damage due to extended stress and sensory overload which requires careful management throughout prolonged engagements.

The functioning battle environment enhances the sensitivity of sensory organs. Battlefield exposure sharpens vision while making ears more sensitive. The sensory organs receive major boosts from stress hormones and adrenaline release. The ability to see in low illumination and recognize details enables the rapid assessment of threats in the environment. Enhanced vision gives soldiers the ability to detect subtle changes in their environment which help them to survive. Soldiers develop the ability to use their eyes and ears to detect threats from enemy activity early on. Hearing faint sounds that indicate danger remains important despite the necessary noise to monitor enemy movements. Enhanced auditory senses are required to make this detection possible. Overstimulation of sensory systems due to continuous, prolonged activation resulted in fatigue or damage that required management to maintain focus for employing enemy countermeasures.

The excessive enhancement of sensory perception in combat situations proves to be counter-productive. Amplified auditory and visual senses help detect potential threats but can overwhelm the brain with excess sensory information which results in sensory overstimulation. Cognitive functions required for sound decision-making become impaired which leads to a greater chance of errors during combat situations. Extended periods of exposure to intense sensory inputs can result in sensory overstimulation which eventually decreases auditory and visual sensitivity and undermines response efficiency during continuous operations. Extended stress exposure can produce lasting harm in the form of hearing loss and disorders from extreme stress situations. The creation of systems to regulate and balance acute sensory amplifications is necessary because they can result in destructive long-term effects. The compelling nature of this clarity sets a standard for full vitality that drives people to seek a return to that condition.

The Double-Edged Sword: Short-Term Gains and Long-Term Costs

These physiological adaptations, which save lives, also have significant downsides. While warfighters perform at maximum efficiency through these mechanisms, they also experience severe health consequences. When the cardiovascular system operates in continuous overdrive, it faces long-term damage risks, including hypertension and heart disease. The respiratory system faces problems such as chronic hyperventilation and decreased lung capacity when subjected to intense activity over a long period of time.

Extended cortisol release disrupts metabolic balance and heightens the possibility of developing insulin resistance and chronic fatigue syndrome. Sensory enhancements that start as advantages can eventually cause overwhelming stimulation or permanent conditions such as tinnitus. The body's adaptation to a hyper-aroused state makes normal life seem unbearably dull, which plays a major role in creating combat addiction.

The warfighter, therefore, faces a profound paradox. Combat provides a physiological rush that brings both mental clarity and intense focus, while civilian life after war appears dull and emotionally numb. The difference between combat intensity and civilian life creates a powerful desire to return to war because soldiers seek the exclusive experience of heightened existence.

Conclusion: The Addictive Nature of Combat Chemistry

The complex interaction between hormones and neural signals together, with physical adaptation during combat produces an awe-inspiring performance state that becomes addictive to many. The body launches into a survival mode with fast heartbeats and quickened breaths alongside enhanced energy and sense perception to cope with chaotic situations. The same biological responses that provide survival advantages in battle also create a sense of focus and clarity that warfighters find themselves missing after returning from combat.

This chapter examines combat physiology not only as a scientific analysis, but also as an explanation of the underlying attraction that pulls numerous individuals back to war. The powerful draw of this experience lies in its essence where every moment reaches peak purity, instinct takes control over conscious thought, and body and mind unite as a singular entity. By examining this biochemical process, we gain insight into the powerful and persistent need to return and relive that state of intense focus and existence.

Combat experience becomes both an external struggle and an internal fight against our biological nature, which creates permanent changes and a longing for frontline intensity. As we move on to later chapters where the psychological and emotional consequences of these physical changes will be explored, this foundational understanding serves as a crucial reminder: The experience of combat tells a dual narrative that reflects both our physical states and mental processes. The nature of military service attracts many who become dependent on the unambiguous nature that war delivers through its merciless precision.

Chapter 3: The Fog of War

War is destructive and transformative. It embodies contradictions. It is an act of both bravery and cowardice; both imposing and suspending hierarchy. Disorder and rebellion are always indexed by strong interaction when one of the parties is weak and has no option to survive, but to adapt and reject the seemingly obvious line of development. War is a huge factor in psychological transformation. And this transformation is revolutionary in the sense of identity, whether this identity is human, social, historical, or beyond. War metamorphoses human history and provokes profound psychological transformations and deep questions for the astonishing persistence of war throughout human history.

The 'fog of war' describes the uncertainty that often accompanies military operations and its impact on decision-making and behavior in the military field. In warfare, the 'fog of war' encompasses the physical condition of the battlefield and also the psychological effects on soldiers fighting on the front lines or individuals directly involved in combat. What the 'fog of war' highlights are how it clouds clear thinking, hindering the planning and ethics that are essential to military operations. Due to the lack of complete knowledge in chaotic situations, the cognitive and emotional judgment of decision-makers can deviate from their intended plans, leading them into unpredictable circumstances. Furthermore, the 'fog of war' creates uncertainty within groups, negatively affecting their psychology and decision-making, which can result in ethical and moral breaches.

The Cognitive Mechanics of War

Cognitive and emotional intelligence are two crucial aspects of human behaviors that are impacted by combat. In stressful combat situations, these are heavily challenged by the 'fog of war,' which makes the environment very uncertain. The main issue is that cognitive intelligence suffers because there is nowhere near enough information, and what is available might not be trustworthy, making it hard to judge risks and results accurately. This problem can lead people away from their strategic goals. Under these chaotic conditions, it is also difficult for people to understand and react to others' feelings and emotions, so emotional intelligence is affected, too. These two key aspects — thinking and understanding emotions — are at risk of being used incorrectly or for poor or morally wrong decisions. The impact of war on perception, memory, and decisions taken by individuals and groups are enormous. Cognitive biases manifest prominently when individuals are exposed to high-stress conditions, as in the case of warfare. Decision-making processes are often a threat to the occurrence of cognitive biases such as the infamous tunnel vision that we hear about in our history books. Tunnel vision restricts the scope of information to the individual, ultimately risking making critical mistakes in decision making. The pressures involved in a life-threatening situation also malfunction and misinterpret time. The protagonist often perceives the current moments to be extended or contracted, hindering their ability to formulate a strategic reply. Such cognitive states are a direct result of warfare and the threats it imposes on individuals and groups alike as they grapple with distorted cognitive processes to view the scenario through the sharpest lens.

Cognitive biases, which are systematic deviations from rationality, are threaded across the warfare scene, significantly shaping the perception and decisions made by the warfighter on the ground. Confirmation bias is a major cognitive bias which leads a person or decision-maker to search and interpret information subjectively to boost the beliefs of the existing score. The Vietnam War allowed military leaders to heavily face the critique of confirmation bias, while the military officials tended to assess the intelligence in a way that underestimated the threat level of the opposition with the support of mythical beliefs regarding the American supremacy. As multiple cognitive biases influence decision-making during war, the framing effect influences decisions significantly, based on how information is presented rather than how valuable the information itself is. As an example, World War II favored the framing effect, which was responsible for determining the civilian casualties as collateral damage during the bombing, eventually reducing the moral bootstrap of people, leading them to treat the human toll during strategic bombings as normal. Having stood in the crater left in Nagasaki, Japan, I can say with certainty that while the case may be argued that dropping the Atomic Bombs may have saved far more lives in the long run, we cannot as humans, ignore the reality of the toll that took on the human soul of global society.

Another cognitive influence that cannot be ignored is a phenomenon known as 'time distortion' which refers to a warrior's ability to perceive the passage of time during stressful situations accurately. Soldiers may perceive that, in certain situations, time speeds up or slows down at random moments, which negatively impacts their ability to make quick tactical decisions on the spot. Additionally, the time distortion related to stress makes it more difficult for soldiers to accurately assess the threats and the opportunities that could come, leaving them making decisions that could be wrong. Time illusions accompany militaristic features and situations, where time could incredibly affect battlefield situations, where another second could change the entire outcome of the conflict.

In combat, one of the most critical and concerning limitations that affects performance is tunnel vision. Tunnel vision is a cognitive bias that is present in most high-pressure situations and can negatively influence the decision-making process and the level of situational awareness that an individual possesses. In such cases, this phenomenon narrows the focus to a single object that he or she obsèrves, while other peripherals of consciousness are neglected. The ability to retain a global picture of the network in warfare situations is unparalleled to any other aspect. Tunnel vision violates one's ability to process information, aiming at the need to address a current threat. As warriors believe they can take other actions when confronted later by a danger, its presence can lead to critical mistakes.

Group Psychology in War

Within the 'fog of war,' group psychology and groupthink manifest in a far stronger way, influencing decision-making. The extreme stress and uncertainty characterize this environment and, consequently, the decision-making process can rely on group thought rather than being critically assessed by individuals. Such collectivization may produce outcomes that are harmful to the ethical dimensions of the military strategy stipulated, due to the acquiescence towards group thinking. This groupthink happens when uncertainty and stress are at their peak and shows how group decisions transcend individual thinking, often overruled by the rationale of the majority. Therefore, encompassing the psychological paradigm because of the 'fog of war,' the individual reasoning is stripped away due to the collectivization of the thought process, while the implications for the ethics of the military strategy are catastrophic.

Group psychology has a major impact on an individual in war scenarios. It often leads the individual to act in line with group behavior and makes an open environment for conformity. This behavior is often driven by the 'us or them' mindset and leads to polarization and eventually dehumanization of the enemy group and the war in general. The case of the Afghan war shows that the deployed military units had a strong sense of unity, allowing members of the same unit to conform and act under obeying the group to impose decisions and direction over the moral reasoning they had over the rightness of their actions. Group dynamics use propaganda to spread a common and popular opinion that demonizes the enemy while imposing their own undisclosed view. All these factors emphasize and confirm that group psychology in war allows for homogeneous decision-making and behavior that promotes a culture of violence and shapes public perception, complicating the resolution of conflicts at the end of hostilities.

In war, the beliefs and values of an individual can be influenced by group or team conformity. The Vietnam War showed how soldiers conformed to the values of fellow soldiers even when they had doubts about the war effort. The fear of losing unity among battalions or platoons, especially under the stress of combat, is a strong force to conform to the group's beliefs rather than individual values or beliefs. Individual judgments of a moral and ethical nature can be discouraged through this group conformity, ultimately leading to actions and decisions that are against personal ideologies. The military underlines the importance of loyalty and unity, creating conformity and discouraging dissent. Many positives and negatives arise from the issues surrounding group conformity, particularly when it comes to an individual's values and beliefs. Understanding how conformity works in a psychological or sociological group is important to the issues faced with morals and ethics in warfare. The primary reason for this statement is due to the tension that exists between one's beliefs and conformity to a group in the context of war and military action. As someone who remembers the events of September 11th, 2001, vividly, I'm drawn back to the Middle Eastern small business owners and members of our communities who were vilified and targeted in the days following the devastating attacks. While individuals are able to understand the difference between the Muslim (or Indian) shop owner down the street from radical terrorists, the group mentality makes that distinction more difficult and makes it easier to dehumanize the enemy so that we can do what needs to be done.

As such, propaganda serves to foster a collective identity and align mass behavior with the objectives of the state during wartime. By creating narratives that support national interests, military and state leaders can unify the populace behind their decisions and actions. As seen during wartime propaganda, framing the enemy as clearly evil, emphasizes the "us versus them" mindset, often removing doubts about behavior and promoting loyalty for the collective good. Not only are leaders able to instill a sense of patriotism and promote action in support of the state approach, but they are also able to effectively justify the moral and ethical implications of war itself by removing the complications of human actions. Propaganda, therefore, serves not only to impact violence or warfare directly but also to help shape national consciousness, providing insight into the post-war environment where reconciliation has been made increasingly more difficult as a result of shaping public opinion.

As we can see, the "us vs them" mentality plays a vital role in deepening conflicts and increasing divisions, as an adversarial position of individuals has a considerable impact on their perception of the challenges they face. Essentially, the "us vs. them" mentality generalized all issues into the simple binaries and encouraged group polarization, which inevitably led to the growing of conflict, distrust, and anger within our own communities. Supporting dehumanization narratives have only reinforced the existing divisions while justifying the aggressive actions against the opponent and denying any possibility to consider them human beings, which created the danger in the case of various conflicts investigated. The oppositional thinking of this kind fuels conflicts further and reduces any opportunities for reconciliation, as it strengthens distrust and continuity of aggression. Thus, the adversarial "us vs. them" mentality complicates the existing conflicts and inclusive separated efforts between the groups, highlighting the destructive impact of the oppositional thinking on the intergroup conflict development and resolution.

The second variable is collective identity, which enhances participation in conflicts because the psychological motivation brings individuals closer to a cause or ideology, which could result in a joint and strong effort from that particular group of people. During the First World War, the nationalist ideology caused many citizens to sign up for the military because their cultural identity brought members together to defend their interests. This was seen on all sides. The collective identity causes psychological motivation in the individuals that form the group, which moves them differently than if they were alone, fulfilling their duty in accordance with the morale created in the collective. All these examples help to understand how the use of the collective identity can support participation of the people in the conflicts, and the involvement in the escalation of the hostilities that these people will take part in. Thus, this is a useful variable for comprehending the psychology of war.

The Moral and Ethical Dimensions of War

One of the hardest parts of the 'fog of war' is the ethical problem faced by people or groups involved in the conflict. During the heat of battle, it's very hard to maintain justice and moral actions as good behavior, because chaos takes over, making it hard for any side to see future threats clearly. Ethical actions become very unclear, and a soldier must choose between following the existing justice system or other principles based on their immediate needs in combat. As a result, everyone involved in the war challenges these moral choices. Also, the way a group thinks collectively can make it hard to see if the actions of soldiers are justified for any side in the war. Decisions made by soldiers as a group can cause moral problems because group pressure can influence a person more strongly than their own moral beliefs. A perfect example of this is the infamous scandal that occurred at the Abu Ghraib military prison during the Iraq war. People who were otherwise normal, morally aware individuals committed horrifically unethical acts in the name of groupthink justice and wanting to conform to peer decisions and expectations.

Ethical compromises and moral dilemmas are inescapable in war settings. They are brought about by the actual engagement in combat, and succinctly by how society impacts lives and its participants in recognizing the calls of war. This is an injury that results from carrying out an act that violates the morals and principles that define an individual. This has been found to have adverse impacts on the mental well-being of military members who are reintegrating back to society in terms of their health. Ethical compromises in the justification of actions taken during warfare are socially founded. These are actions that could not be justified in a peacetime situation but are taken in existential wars and being made to seem acceptable.

Wartime environments frequently do require moral compromises in which it becomes possible for important military objectives to take precedence over ethical ones, with operators in conflict prioritizing actions that result in a more secure environment and objectives over moral culpability. How do you reconcile the moral damage knowing you took the life of another person, when despite any natural draw to combat, we are also ingrained to believe in the value of life? An example of this moral flexibility is destroying civilian areas solely for the strategic imperatives observed during wartime operations. Individuals involved in military operations may routinely find themselves in the situation of having to rationalize actions that oppose their ethical convictions while addressing the psychological requirement for these moral compromises. Such situations highlight the need for individuals and societies at large to traverse the ripple affected moral principles and threatening survival to coexist in the tumultuous path of warfare.

At the collective level, the injuries caused by war also permeate society, leading to moral injury that can be transferred from generation to generation. An example of this can be traced back to the Lebanese wars, wherein the collective society was forced into legitimizing violence and destruction of places and individuals. This led to the collective loss of hope and the ethical disarray brought upon by people due to the impact of the war. These experiences are then reflected backward into time through the descendants of the individuals who lived through it, as the psychosocial implications of the society-level moral injury were passed down to the next generation, fishing for societal injuries that can be used as a currency of post-conflict moral persistence.

The Societal Impact of War

War has undeniable impacts on entire collectivities. Notably, war can significantly redefine identity patterns and social morphology across large geographical expanses. Cultural effects are manifested in the discontinuances of former cultural development patterns, as wars disrupt processes whereby values or ways of living were transferred from one generation or society to the next. Entire collectivities may be displaced, decimating previously existing social networks and levels of organic solidarity. As we've seen time and time again, the psychological effects of a highly armed societal conflict extend throughout entire collectivities in the form of trauma or intractable mental health problems. Such influence illustrates the duality of wars, on the one hand prioritizing the immediate interests of survival for identified groups of people and, on the other hand, redefining social constructs.

A whole generation of the population suffers from intergenerational trauma because of war. Going back to the Lebanese Civil War example, older generation events translated into a lack of psychological development for their mentees. That trauma translates into rising levels of anxiety, depression, and other varieties of psychopathologies. People affected by previous war events develop long-term trauma. Its strong effect makes it impossible for whole communities to reshape their embedded cultural and social identities in a post-war context. Here in the United States, we saw this with the drawn-out Global War on Terror. By the end of the wars in Iraq and Afghanistan, young troops were being sent overseas to fight a battle that began before they were even born. We saw parents and grown children fighting side by side in a conflict that crossed generations. Even now, it is difficult to find someone who was not affected by a conflict that spanned across three decades. This long-lasting trauma needs to be exposed and understood in order to implement strategies of health and well-being that take into account the psychological health of children's parents who lived through wars and their descendants. Therefore, it is possible to describe a big pattern of effects that war generates in humans, rather than just a psychological illness framed within scientific epidemiology from war. Even now, I believe we are remiss in allowing the propaganda of nations to continue to refer to it as the "Global War on Terror" instead of calling it what it really was...World

War III. While many argue that calling it WW III is unjustified, there is a logic behind it. Nearly every developed nation with an organized military played some role in this war, which is something that cannot be said about any other war in contemporary history.

The normalization of violence is a key psychological and social consequence resulting from living in a constant state of conflict. The individual and communal experiences and understanding of violence change significantly within society. A society that has experienced long-term conflict and violence may normalize similar acts, causing societal norms and pressures to work less effectively and violently toward aggression. The community becomes desensitized from acts of violence and corruption, which may result in society considering these acts as the only means to survive peacefully or controlling the situation. For example, in areas affected by long-term conflict, such as the Middle East, evidence suggests that normalized violence infiltrated daily life and social behavior among societies affected by the situation. As it normalized and continued, similar acts also become embedded in the society making it extremely difficult to build and maintain peaceful and stable social cohesion.

Collective grief following conflict of war can affect national identity as well, as society goes through the process of mourning and recovery. Losing lives and heritage in war can be significant for national understanding, and the community can go through this process of mourning together. National identity can be further united from collective sorrow experiences when those losses bring people together and become a part of their community. Sorrow after losses can result in significant threats or benefits for the nation, which is a crucial reason for collective grief management as an important intervention to focus on healing from past losses' wounds together and build a national community.

War is a powerful agent of cultural change and societal adaptation. Society always resorts to adaptation when the environment is applying too much pressure. In countries characterized by years of wars, certain cultural norms will face challenges and remain modified because of adaptation. Such changes directly impact the cultural paradigms within society, and not only this, but also normalize situations that used to be considered out of the ordinary. During war, the proportions of social norms are those that define the roles of men and women, children, work groups, and other associations, as the society that faces war leaves its conditions recognizable and the people must adapt constantly to survive within them.

The impact of long-term warfare brings extensive psychological alterations to societies. Coping strategies that involve adaptation and resilience have become an indispensable means of survival to people in the areas where many conflicts occurred. Such networks often become essential psychological shelters for those exposed to constant adversities because they allow most individuals to maintain normalcy, despite external constraints. The need to adapt also contributed to the emergence and development of cultural and social practices that aim to emphasize the importance of group solidarity and emotional well-being, which demonstrates the influential ability of communities to adapt and survive in the conditions under which their normalcy has been abandoned due to ongoing challenges.

War plays a critical role in establishing certain values and priorities in human society since it leads to the cultural revaluation of certain important aspects in daily life. In times of war, certain values become more important than others. Human society can make substantial changes to its priority list when facing a war. For example, the long-standing conflict in the Middle East has led society to change its values to a priority in which women take part to support the war or work for the family to sustain the family. These values that endure after the conflict can completely rotate the priorities of a certain sector in human society, thanks to the war experience that can trigger a revolution in certain accepted aspects.

Conflict-induced societal transformations have both direct and indirect implications for both future conflict and peacebuilding. Societies engulfed in conflict for a long period experience a shift in their cultural narratives that can potentially continue the violence, as the cultural norms that result from a war often continue to be in place, shaping perceptions and strategies in new wars. For instance, the adaptation to violence and ongoing conflict in places like the Middle East, where an entire culture seemingly promotes warfare and violence, complicates the establishment of peace-building efforts, as these beliefs are deeply rooted. Another implication would be the collective identity formed due to warfare, which emphasizes the importance of unity against its enemies; however, it also limits acceptance and reconciliation, as divided identities play a critical role in post-conflict societal changes. Such implications reveal the importance of breaking deep-rooted cultural and societal transformation, as there remain significant repercussions due to warfare. Targeted interventions are necessary to remove the enemy identity and develop a supporting climate for long-lasting peace.

Conclusion

The ambivalence of human ability to destroy and to endure, manifested in war, is seen as a paradox. War has a capacity to inflict devastating blows upon humanity; it can destroy societies and civilizations, eradicate cultures and peoples, and bring about human sufferings of an unprecedented scale. On the other hand, war reveals the staggering ability of men and humanity to endure pain, survive destruction, and take on the difficult task of rebuilding what could seemingly have been irretrievably lost. These contrasting facets of the human condition raise some urgent, yet troubling questions about the role of war in our history; why does humanity continue to inflict upon itself the calamity of war as a form of resolving conflict, when it has shown its capacity, however painful, to heal, to reconcile and to live together, and what does it mean for our future?

Chapter 4: The Strength of the Wolf Is the Pack

The psychological domain encompasses essential warfighter traits, including resilience, stress management, and adaptation ability. The most essential yet simple inquiry is why people choose to expose themselves to war's damaging dynamics on their own initiative? Understanding the person's psyche remains a fundamental aspect when facing other elements that either mitigate or intensify the psychological impact on individuals living through these extreme experiences. War creates exhausting stress and effects which lead people through multiple psychological changes during and after specific experiences that later influence their mental state and actions.

Emotional Resilience and Coping Mechanisms

Warfighters receive extensive theoretical training for psychological resilience development, which is strengthened through practical skill exercises during field operations. Emotion regulatory self-reflection training enhances both knowledge and execution of resilience coping methods to improve performance levels in extreme environments. Over the years, the internal health of soldiers during combat operations has improved thanks to emotional resilience and coping strategies. The formation of strong bonds between soldiers establishes an essential support system for both resilience coping strategies and related emotion regulation techniques. Indeed, nobody can understand the reality of that world who hasn't been there. During warfare and combat, the warm support from trust and comrade bonding functions as a safety net alongside coping strategies for mental injuries. Engaging in humor during tense situations acts as a coping mechanism that shifts focus away from actual warfare, which helps soldiers to handle complex stress and enhances both their morale and performance. Resilience coping strategies promote both emotional resilience development and performance cohesion improvement. Contemporary psychiatry demonstrates that emotional self-reflection training strengthens soldiers' performance assurance, which helps them manage their environment better.

Soldiers use emotional desensitization as a common coping mechanism to handle the psychological demands of warfare. Warfighters experience a reduced sensitivity to emotional triggers that lets them endure distressing emotional stimuli without being overwhelmed by their emotions. Military training programs support emotional desensitization because they condition warriors to maintain their composure and demonstrate mental toughness when facing stressful situations. Training enables soldiers to handle particular emotional triggers while establishing unit compliance and building psychological resilience in each member.

The development of emotional resilience in warfighters depends heavily on camaraderie as it builds a support system based on mutual trust and shared experiences. The bond between soldiers creates a psychological shield which reduces stress through a supportive environment where they can depend on each other amidst combat-induced emotional challenges. The feeling of connectedness combined with the confidence that they will not face challenges by themselves helps warfighters to better manage severe mental stress. Researchers have determined that military training programs which utilize emotion regulation strategies to improve group unity, strengthen this connection and result in better operational outcomes and psychological toughness for military units. The power gained from interpersonal connections becomes vital for soldier morale during traumatic events because it strengthens both personal and group resilience in military operations. In his autobiographical account of the fierce fighting in the Battle of Okinawa during World War II, E.B. Sledge wrote of camaraderie's influence in motivating others forward during combat.

> *"I lowered my head and gritted my teeth as the machine-gun slugs snapped and zipped around us. I expected to get hit. So did the others. I wasn't being brave, but Redifer was, and I would rather take my chances than be yellow in the face of his risks to screen us. If he got hit while I was cringing in safety, I knew it would haunt me the rest of my life – that is, if I lived much longer, which seemed more and more unlikely every day."*
>
> - E.B. Sledge, With the Old Breed

During my time in the Marines, six decades later, I learned what this brotherhood meant. On May 8th, 2005 (Mother's Day) 2nd platoon, Kilo company, 3rd Battalion, 3rd Marine Regiment was ambushed by insurgents in the Alishang Valley, located in the Lagman province of northwest Afghanistan. After hours of fighting, my platoon accounted for 15 confirmed enemy KIA and another six confirmed enemy WIA. This major engagement was not without our own losses, unfortunately, as 2nd platoon suffered three wounded and two killed in action, Corporal (promoted posthumously) Nicholas "Nick" Kirven and Corporal Richard "Ricky" Schoener. First Lieutenant Stephen Boada would later be awarded the Silver Star for his actions that day, while Nick and Ricky were both posthumously awarded the Bronze Star with V device for valor. After the initial battle was over and the damage was assessed, we needed a way out of the valley. Helicopters were not an option as the risk of RPGs was too great and the roads were unpassable by vehicles, meaning that ground transport was not an option. Eventually, the decision was made that we would carry out our fallen Marines over the several miles of dangerous, mountainous terrain. The Marines of 2nd platoon were tired. We were hungry, dehydrated, and mentally and emotionally exhausted. But we still had a mission to do. While continuing to receive air support from 2 A-10 Warthogs and 2 AC-130 gunships, we made our way out of the valley under the cover of darkness. Not a single Marine was willing to let down his fallen brothers or risk allowing their own exhaustion to endanger those still standing. During the trek, several small enemy ambushes were faced. The roar of gunships lit up the mountainsides above us, telling us that there were even more ambushes that we were saved from thanks to those courageous pilots. As the sun rose over the valley, we reached the mouth, where the rest of Kilo company was waiting for us. Over the course of my four deployments (three of which were combat deployments to the most dangerous places on the planet), I have learned that regardless of how long you spend, every warrior has their defining moment that embodies an entire war for them. For those of us in Kilo 2, that was our moment. Regardless of what other battles we faced, this is the one that lives with us forever.

It was the first moment we realized we were not invincible war machines.

After having spent two combat tours in Afghanistan and returning to civilian life for two years, the Marine Corps decided they needed me to return for one more combat deployment. Units headed back to Afghanistan desperately sought experienced operators to fill holes in their ranks. Having already faced the horrors of combat in the mountains and valleys of Afghanistan and knowing that Iraq was beginning to wind down, making it (relatively) safer, I made some calls and got assigned to a unit deploying to Iraq instead of Afghanistan. While it was still an active war zone, I knew that the risks (at that time) were lower in Iraq than they were in Afghanistan, though hardly non-existent. About halfway through my deployment, my dear friend and brother Ryan Lane was killed in action in the mountains of Afghanistan. This decimated me probably more than any of the other brothers I had lost. Not because I was any closer to him than the others, and not because the other losses I had faced were any less tragic, but because of the inherent guilt I felt over knowing that this brother I had fought next to on countless other occasions, returned to a battlefield that I worked to not go back to. Over the years, people have tried to help with that guilt by pointing out that I still chose to return to combat. I was still in danger, just in another country. While that may be true; as warriors, it's still hard not to internalize that guilt. Had I been in Afghanistan, I most likely would have been with a different unit, unable to help, but maybe I could have. That is an emotional injury that never really goes away.

Humor turns into a vital psychological coping mechanism during combat that helps soldiers manage stress by offering relief from the constant tension of military operations. Soldiers leverage comedic relief in intense situations as a means to reduce stress and sustain their morale. I vividly remember a nighttime attack on a forward outpost I was assigned to during one of my deployments to Afghanistan, one of the Marines in my unit (who will remain unnamed to avoid embarrassment now that we are all much older with careers and children), engaging in the fight wearing nothing but his protective flak jacket and helmet. The hilarity of the moment helped diffuse the uncertainty and chaos of the firefight. Instead of being a distraction, it helped relieve the tension that would have otherwise clouded our thinking in that moment. Through shared laughter, soldiers develop unity, which enables them to connect through common experiences, while reducing the stress of combat. Soldiers gain mental strength from communal laughter which simultaneously provides them with momentary relief from combat horrors and helps them face adversity. By incorporating humor into everyday military activities, soldiers build emotional resilience, which helps them regain a sense of normalcy amidst surrounding chaos.

Ethical and Emotional Weight of War

The major issue of moral injury highlights the significant effects of war on soldiers' ethical and emotional frameworks and their mental health. While the issue of moral injury was mentioned briefly in the previous chapter, we cannot ignore its effects when we talk about the psychology of the individual warfighter. Moral injury arises when individuals engage in or observe actions that breach their core moral principles, leading to feelings of guilt and shame. Research has connected moral injury with mental health issues during deployment yet notes this field remains under-researched while featuring psychological and spiritual dimensions. When actions contradict personal values, individuals face significant challenges in mentally justifying these damaging behaviors. Soldiers face mental health challenges when they experience ethical conflicts between duty and ethics during combat, which indicates an urgent need for interventions that address psychological and moral-spiritual harm.

Warfighters must face moral injury as a psychological threat because it creates intense emotional suffering accompanied by guilt. Warriors experience moral injury when they perform actions which violate their ethical beliefs. The opposition between moral concepts and real-world events generates a distortion between ethical principles and actual occurrences which ultimately precipitates war. Warfighters suffer from moral injury because they experience ongoing psychological distress through feelings of guilt and shame alongside spiritual pain which complicates their mental well-being. Internal conflict in warfighters produces serious psychological outcomes that subsequently destroy their ability to discern right from wrong following the end of combat operations. Therapeutic intervention for warfighters requires consideration of moral injury which manifests through psychological and spiritual dimensions.

The psychological and emotional damage warfighters experience when they struggle to align their wartime experiences with their personal values and beliefs can result in psychological harm. Researcher Victoria Williamson and colleagues describe moral injury as soldiers experiencing profound shame and guilt which often leads to psychological damage. Soldiers who struggle with this reconciliation risk developing psychological injuries from their internalized emotions and feelings over time. War veterans must receive therapeutic interventions that support both spiritual healing and psychological recovery to enable them to manage their experiences and reorient their behaviors to align with their core values. Their moral beliefs become restored through this process.

In what can seem almost like a contradiction, this struggle with moral injury can be a major source of desire to return to combat. The conflict in these actions occurs because they are trying to align what they did with what they believe based on society standards. When a warrior returns to combat, they resubmerge themselves into a world that makes sense to them. Where the lines are less gray and they are surrounded by others that are facing the same moral and ethical dilemmas, bringing a sense of peace and acceptance.

Reintegration Challenges

Ex-warfighters experience significant challenges when moving to civilian life due to mental health struggles and emotional reconnection issues. Post-Traumatic Stress (PTS) encompasses numerous psychological complications which hinder veterans from smoothly transitioning into civilian life. Military veterans experience hyper-vigilance and anxiety as well as re-experiencing their trauma, which makes their transition difficult because of their combat experiences. The persistent struggle veterans face when entering civilian jobs underscores the difficulty of transitioning into civilian work as a crucial step for societal reintegration in a more approachable way. Veterans face barriers in emotional reconnection because their training to hide emotions during battle prevents them from recognizing and connecting with their own feelings. These veterans often struggle to reconnect with other people and face obstacles when trying to integrate into their community through social activities. Therapeutic interventions focusing on resilience and emotional openness should address reintegration issues to help veterans navigate the civilian life transition and improve their post-service life.

When attempting to overcome the challenge of societal reintroduction, soldiers need to use the adaptability they developed in the military to help transition into the civilian world while dealing with severe psychological disorders such as PTS, characterized by anxiety and intrusive thoughts that block their departure from military organizations. Therapy serves as a key method to offer personalized solutions for this mental health issue while enabling soldiers to learn adaptability through resilience practices and emotional exposure. Through resilience-training programs that incorporate emotion regulatory strategies, individuals learn adaptive coping mechanisms, which help soldiers transition to civilian life.

The post-war meaning-making process plays a crucial role in helping warfighters recover and successfully reintegrate into civilian life. 'Meaning making' describes the mental and emotional processes people use to reinterpret experiences so they can integrate traumatic events into their life stories while developing both meaning and purpose in their lives along with post-traumatic growth. For many warfighters who have experienced the primal force of combat, finding meaning and purpose after combat is one of the most difficult fights they will ever face. The difficulty in transitioning from military service to civilian life exists because personal experiences do not match societal expectations. Veterans employ thinking and storytelling as therapeutic methods to reconcile their past actions with their future objectives and personal growth. When veterans share their stories about overcoming difficult military experiences, they develop resilience, which facilitates their civilian life transition, and serves as emotional healing.

The adjustment to civilian life poses emotional challenges for warfighters who must transition from a high-functioning military emotional environment to a lower-functioning civilian emotional environment. Building personal relationships within civilian life demands open emotional expression, which starkly contrasts military experiences. Warfighters face difficulties in relating to their families and peers because their realization process creates a barrier which might lead them to forget how to reconnect when needed. When warfighters fail to form emotional bonds with their family members, they develop attachment problems and begin to withdraw socially. Stimulating emotional responses in warfighters through specific interventions will assist them in overcoming emotional challenges and rebuilding relationships with those close to them.

The Paradox of the Warfighter

The warfighter paradox emerges because we are expected to exhibit heroism while remaining vulnerable. Warfighters receive hero status because they demonstrate resilience when facing challenges. Heroic actions occasionally create an impression that warfighters possess unbeatable strength. Despite their heroism, warfighters experience moments of vulnerability and severe mental health challenges. Warfighters become vulnerable through exposure to combat-related experiences. It affects their cognitive functions as well.

Heroic actions can sometimes damage memory function and impair decision-making abilities. Warfighters face continuous challenges because they must balance their invincible image with their emotional fragility which turns them into a perpetual battlefield.

The warfighter already embodies both hero and victim characteristics within themselves. The dual way of perception forms a psychological setting that is both conflicting and complex. Soldiers demonstrate heroic achievements in combat yet endure hidden vulnerability and risk factors as they maintain mental stability while battling emotional instability. There are deployment demands. The fearless and altruistic actions soldiers display during warfare build an impression of faultlessness which society sees as a demonstration of power. The warfighter identity involves both a deliberate construction of power through self-transformation and an unconscious acceptance of this state because fighters find joy in combat. Reconciling this inconsistency is difficult when returning to the civilian world, which is why many feel more comfortable returning to the world of combat. There is great comfort in knowing those around you are experiencing the same paradox and have no unrealistic expectations of you.

The resilience-vulnerability dichotomy represents both the psychological struggles and mental health intricacies of warfighters. Warfighters need to develop resilience to operate under stressful conditions while also facing psychological vulnerabilities that chronic stress can produce. Intense warzone exposure limits higher order cognitive processing even when resilience measures are taken, which reveals mental health challenges in warfighters. The warfighter faces unique demands as the strength-vulnerability differential creates a mental health dichotomy that requires specialized resilience measures. To strengthen battlefield psychological resilience, we need to understand how vulnerability affects resilience and apply measures that enhance resilience while reducing psychological health stressors and maintaining optimal mental health in warfighters.

Broader Societal Questions

The primary consideration when addressing societal questions should be our responsibility to provide care for soldiers after their return from combat. The society has an obligation to monitor the physical health and psychological and emotional states of soldiers after they return from combat. Society must maintain mental health support for soldiers beyond their immediate return from duty because they face enormous psychological threats and perilous outcomes. The thrill of performing military duties during war might be a significant reason why a soldier wants to return to battle. When examining why a soldier continually chooses to return to war, one uncovers elements like adrenaline rushes, and purposeful meaning, along with meaningful military relationships. War is instinct. Society needs to maintain its efforts to comprehend the intricate relationship between natural human attraction to normalcy and nature, while simultaneously creating appropriate social supports and guidance systems. The society needs to maintain ongoing discussions about how to manage its approach to understanding and handling soldiers' temptations to return to war as they reintegrate into civilian life.

A society's obligations towards warfighters during their transition back to civilian existence play a critical role in their successful reintegration into civilian life. The full range of needs that soldiers encounter when transitioning from military service to civilian life is encompassed in society's approach to post-combat support. Support programs that focus on Post-Traumatic Stress (PTS) help veterans manage symptoms like hyper-vigilance and anxiety. Ongoing societal support for veterans' mental health requirements demonstrates an essential commitment needed after they return from combat. Warfighters need essential continued care support during their transition back into civilian society. By completing their post-combat duties, society can honor warfighters' sacrifices and help support their reintegration and recovery.

Warfighters return to battle because psychological and emotional aspects that originate from the war experience drive their motivations. Warfighting creates an addictive condition because the intense adrenaline rush experienced during battle cannot be matched by civilian life activities, which drives soldiers back to combat zones. The power and control that others hold can also be a motivating factor in such an instance: The military's structured hierarchy provides clear authority, which is difficult for individuals to attain within their complex civilian lives. A sense of camaraderie exists within military units, which connects to both military hierarchy principles and the brotherhood felt inside military units, while the spirit of unit togetherness serves as one reason why individuals return to war. These elements reveal war's instinctive nature and its ability to draw warfighters back into its legendary allure while the societal understanding of what drives this behavior remains insufficient.

Conclusion

Warfighters who serve in the military are often burdened with complex psychological, emotional, and ethical challenges that embody the layers of military work and continuous effects to individuals. Cognitive strain and combat stress are key forms of pathology implicated in the experience of the warfighter's lifestyle as it impacts their operations through impaired judgment and increased risk. In terms of emotion, bonding through appreciation, humor, and other tactics act as important detractors; however, moral injury due to major combat ethical dilemmas also entails ethical effects that persuade moral injury. When the warfighter returns home, their mental and emotional experience provides structural problems for societal reintegration as PTS, societal adaptation, and identity reform demands are present. Their experience encourages public efforts toward understanding military work and its ethical effects on the veterans as societal obligation; therefore, more efforts must be accounted for in providing, framing, and developing efforts that seek to recognize warfighter sacrifices through societal support and attention to their integrated recovery.

Chapter 5: The Tribe We Leave Behind

The harsh environment of combat creates unique bonds between warfighters that civilian settings cannot duplicate. The connection among those who face combat together becomes more than friendship. It evolves into a sacred agreement based on survival, loyalty, and trust, which unites them as a purpose-driven tribe. The tribe dissolves after the battle ends but leaves behind an intense feeling of loss. The loss of military camaraderie stands out as the most destabilizing experience for many veterans over missing their missions or the adrenaline rush. The tribe stands as much more than a friendship network because it represents home and family together with identity.

This chapter will examine those bonds and explain how the connection between warriors is just as much of a reason for returning to war as any physiological or psychological reaction. The intense bonds that develop from shared danger and sacrifice become fundamental aspects of warfighting psychology.

Building the Tribe: Bonds Forged in Fire

Military life is filled with extreme and daunting circumstances, and the battlefields unite warfighters in a way that only they can understand. Their experiences together help them survive as they build a tribe full of trust and loyalty. This bond gives them a sense of belonging, and they have no choice but to depend on the fellow soldier's trusted decision-making skills to help them through life-threatening situations. While this bond is, of course, extraordinarily strong, it is also very immediate in the sense that it fosters a dependence in the moment. Camaraderie, in this light, becomes an integral part of resilience as it offers emotional protection against adverse events from outside forces.

Moreover, certain instances of how shared experience of combat further deepens the critical relationship with interdependence: In combat, the life of every soldier is dependent on the actions of the people who are alongside him/her. This builds a strong level of trust in the lives of each other. Beyond tactical immediate sets, the level of interdependence grows even further, as soldiers cope together with the stresses and traumas of combat. The interdependence leads to a strong member of a community, where the loss of one member is considered the loss of every member. This heightened level of interdependence helps strengthen loyalty and solidarity. As a result, the level of interdependence and shared experience of hardship help develop a bond which often transcends the similar levels of connection and depth developed in the friendships formed in civilian life, sometimes lingering even further.

In the quiet moments between engagements, my team and I would share stories, jokes, and sometimes even silence. Often, we would crowd around someone's laptop and binge watch DVDs of TV shows like Smallville or The OC (did you assume we'd be watching war movies?). Those times weren't just about passing the hours; they were about grounding ourselves in something human amidst the chaos. In those moments, I wasn't just another Marine, I was a brother in a family. This family became closer than my own real family. They became my anchor, and knowing they were by my side gave me the strength to face whatever came next. Everything was done in groups. While this is generally encouraged for safety reasons, we found that we genuinely wanted to be with our brothers when going to the gym or sitting down to eat. Others outside our tribes were at best ignored, at worst mocked for not being one of us. Camaraderie, in this light, becomes an integral part of resilience as it offers emotional protection against adverse events from outside forces. This sense of shared purpose that comes with belonging is rarely found amongst the civilian counterparts, creating a bond that makes troops stick together.

Engaging in life-threatening situations together creates a vulnerability among warfighters, which makes trust and dependence necessary elements for these types of connections. The extreme stressors related to combat generate an emotional and psychological demand on warfighters where resilience is a must. The implication of experience of collective threat stimulates a unitary feeling and a collective identity among the soldiers, who must rely on each other for survival. Therefore, the coexistence of common experience and dependency is an integral component in strengthening the bonds, a distinctive strength that is often carried into their lives after service. After nearly two decades, my Marine brothers and I still have a bond that cannot be explained to or understood by those who were not there. We have been there for each other through the hardships of civilian life in the same way we were there for each other in combat. We faced natural disasters, deaths, divorces, and countless other tragedies together.

The Void Left Behind

The move from military life to civilian life results in unavoidable separation from one's military tribe. Soldiers scatter into different directions once their missions conclude and the habitual brotherhood rituals gradually vanish. People who don't understand the experience underestimate this loss while warfighters experience it as a fundamental part of themselves slipping away. This type of loss is not simply a transition from a highly structured military environment to a more self-sufficient civilian lifestyle, but rather a feeling of deep longing for lost connectivity, for the tribe and its unparalleled sense of unity, which they naturally don't find in any civilian experience.

After my military service ended, I desperately sought that feeling of connection but found that it did not exist in the places I searched. My attempts to build connections at social gatherings resulted in superficial conversations, and the relationships lacked the depth and trust characteristic of my former tribe. The civilian population could not grasp the abbreviated communication style or the dark humor that had become integral to my existence while I struggled to find words to describe it. In college, I found the unique bond and connection felt during combat was impossible to replicate. During that time, the strongest friendships I made were with other veterans who could relate to the things I'd seen and done, the language I spoke, and withdrawals that come from missing the combat high. This disconnect can really only be described as a 'missing piece of my soul' and resulted in years of alcohol dependence and superficial relationships built mainly on sex. It wasn't until much later, when I truly began to delve into the psychology of what I'd been through and what I would need to do in order to survive, that I began to heal and grow.

The absence of tribal connection affects both emotional and psychological aspects of well-being. The team-based sense of purpose and belonging one experiences during a shared mission proves extremely hard to replicate once returning to civilian life. Veterans who experience this loss often develop feelings of isolation combined with restlessness. The lack of tribal connection creates a deep divide that demands both time and deliberate effort to overcome.

The Call to Return

Warfighters tend to return to combat because they find connection in the camaraderie of their group rather than because of the violent nature of war. While the physical and psychological effects of war are indeed addictive, the battlefield also provides a distinct clarity of connection that civilians normally cannot experience. Combat eliminates pretense and hidden motives to reveal pure and direct camaraderie. Warfighters often refer to the battlefield as a place with clear roles, where individuals become part of an identity that prioritizes the collective over the individual. The sense of belonging that one derives from combat is rooted in the importance of day-to-day survival and shared mission, granting then a psychological fortitude that lacks in everyday, civilian life. Those who have left combat and the military service deeply miss the feeling of belonging and understanding others without needing explanations.

Stepping back onto the battlefield in 2008 after a two-year absence brought me profound relief. The tribe was what I desired deeply, rather than the adrenaline rush or dangerous situations. The opportunity to work alongside my team while sharing our objectives and dangers revitalized my sense of purpose. As I mentioned previously, being away from combat was like a missing piece of my soul. When I returned, that part of my soul that I had been missing was finally restored. But that experience also taught me a critical lesson: The tribe remains irreplaceable, yet its impact can persist beyond its existence.

At the end of that 4th deployment, I was faced with a difficult choice. I had once again found my refuge, yet I had to decide whether to stay or get out once again. Ultimately, I made the decision to once again step away due medical issues that I knew would soon start catching up to me. But this time, I had a plan. I wasn't thrusting myself back into the world without purpose. I enrolled in school and began looking for ways to continue to serve beyond the battlefield.

Rebuilding the Tribe in Civilian Life

The essence of combat bonds lies in trust and loyalty, which we can transfer to civilian life even if the bonds themselves cannot be replicated. Veterans find the solution to their loss of tribal connection by reinterpreting their experiences instead of searching for a replacement. The translation of the tribal nature of the battlefield into the civilian nature of life requires adaptive tactics and innovative systems for community development. Veterans successfully build networks that emphasize the values of their tribe by creating subgroups of interest, where loyal support and joint goals are more important than anything else. Access to veteran support organizations and the famous 'small groups' system can play a significant role in solving this problem - the group of veterans jointly creates a communication space where they can safely share and discuss their life experiences and future plans. Also, performing the function of volunteers and participating in social projects allows the former military to express their duty in a socially useful way, which contributes to the preservation of 'team spirit' in life after the service, inherent in the military ethos. Creating the tribes of mutual understanding and shared values allows veterans to repeat the tribal nature of their service and extend this boundless resource of powers for further life.

Through my personal experience I learned that creating meaningful relationships demands deliberate action. I joined veterans' groups where I could meet people who spoke the same military language and shared military life rhythms. Through my coaching work I started building connections by establishing environments that allowed trust and authenticity to develop. I discovered that the spirit of the tribe extended beyond the battlefield to encompass any community that valued mutual respect and collective objectives.

The process of rebuilding the tribe requires the formation of fresh connections that respect its foundational spirit. Veterans can preserve tribal values through mentorship activities, volunteer work or just their presence to help maintain its legacy.

The Lasting Legacy of the Tribe

Friendships created during combat persist after hostilities have ceased. These bonds become integral to our identities, influencing our perspectives on relationships, loyalty, and shared purpose. The bonds formed during combat create permanent imprints that continue to affect warfighters throughout their post-battle lives. Although the bonds of the tribe fade and the common mission ends its course, the teachings from that intense experience remain active and shape all future relationships and decisions while guiding efforts to reconnect with the world.

The tribe establishes its enduring legacy by teaching how loyalty reaches its deepest essence. When soldiers fight together, loyalty evolves beyond duty to become an automatic response. Your comrades place their lives in your hands, and you return their trust by risking yours. The bond formed in such circumstances remains indelibly etched in memory. Through this experience, warfighters learn about profound human commitment, which emerges from actions and sacrifices rather than spoken words or gestures. You understand that leaving someone behind is never an option regardless of the cost. The lessons learned during military service continue to guide you even after you've stopped serving as a soldier. It appears through how you support your family members and friends, as well as aid strangers who require help. A silent promise exists which assures others that my support is reliable. "No man left behind" isn't just a saying, it is a commitment.

The tribe's second key element is its legacy of shared purpose. Every tribe member fulfills a specific function during battle. The mission's success depends on everyone's contributions, regardless of their size. The shared purpose in the group develops a sense of meaning that provides fulfillment and stability. Warfighters continue to carry their purpose forward after disbanding by seeking new opportunities to make a difference and establish leadership.

I've experienced this personally. My military departure left me feeling lost as I searched for civilian missions that offered the same clear importance as my combat experiences. Assisting veterans with their transition showed me that my purpose remained intact but transformed into a new purpose. The tribe showed me how to serve others, which then established the foundation for my life after military service. Finding purpose beyond combat while still serving those who serve has led to a sense of strength and continuity within the tribe I built.

The tribe's legacy includes an appreciation for resilience and collective strength. During combat's most challenging moments, it is typically the tribe that sustains you both physically and emotionally. Acts of connection like encouraging words or shared laughter prove you are never completely isolated. The sense of shared resilience transforms into a continual source of strength for the future. The training demonstrates to combat soldiers they can withstand extreme challenges because they have experienced the power of human connection during dire situations.

This enduring legacy of resilience serves as my foundation of strength throughout every aspect of my life. It pushes me forward whenever civilian life seems too difficult to handle. This resilience enables me to tackle life's obstacles with the steadfast determination I had as a Marine. This legacy drives my work to guide others toward discovering their personal resilience while teaching them about their hidden strength.

At its core, the tribe's legacy revolves around memory. The identities formed by the people we served alongside become integral parts of who we are as we carry their faces and stories forward. They represent the essence of our unfiltered and genuine selves. They remind us of our human essence and motivate us to pay tribute to those who have left the journey. The legacy of memory serves as both a weight we carry and a blessing we hold. The legacy connects us with our history while guiding our future actions to embody the tribal values.

Our former tribe persists within us even after we have moved on. The lessons we learn and bonds we make within a tribe interweave with our spirit to become part of our very identity. We must carry forward the tribe's essence into our future lives by applying its teachings to create meaningful relationships and communities. This approach allows us to sustain tribal heritage which becomes part of our lives and extends into the lives of others whom we impact.

The brotherhood forged in combat lasts forever and is a greater bond than most can understand. Even after leaving the military, my brothers have been there for me, and I for them. They have helped me through my struggles with alcohol and sobriety, through my divorce and custody battle, through the shared grief of losing brothers to suicide, and with the pressing struggles of forging a new path. We are connected at the soul and knowing this connection will always be there has given me strength in my times of weakness and confidence in my moments of doubt.

Conclusion

The tribe we leave behind forms an essential part of the warfighter's identity and serves as one of the strongest explanations for why combat maintains such an intense hold on those who survive it. The intense bonds formed through shared danger and sacrifice become integral to the psychological aspects of military combat. Beyond camaraderie, these relationships represent essential components for survival and purpose in the warfighter's life. The loss of those important relationships creates an intense emptiness that many soldiers find challenging to overcome.

Warfighters owe their compulsion to return to combat to their tribal bonds. Beyond adrenaline and mission objectives lies a deeper necessity; it's the unique clarity and connection only tribal bonds can offer. During battlefield turmoil, the tribe serves as a stabilizing force that brings definite understanding in times of doubt. The profound bonds created through shared vulnerability and trust become so deeply embedded that civilian life afterward appears unanchored and superficial by comparison. The desire to return stems from the longing for tribal bonds which provide raw authenticity and meaningful connections that make everything feel real and significant.

When we move beyond the tribe, we leave it behind, yet it remains within us. The lessons and values learned stay with us and guide our civilian lives while the spirit of our military experience remains a fundamental influence. The framework that enables civilians to construct meaning stems from the loyalty, trust, and shared purpose which characterized those bonds. Veterans face the task of transferring their tribal spirit into fresh endeavors and relationships instead of attempting to recreate the former combat team. Through such actions we pay tribute to the tribe's legacy while discovering methods to transition from warfighter to civilian life.

This is the duality at the heart of the warfighter's journey: Service members forge strong tribal bonds while remaining obligated to establish meaningful civilian lives. The tribe embodies our happiest moments during combat while also becoming the most difficult aspect to leave as we transition to civilian life. The experience stands as both a testament to human connection during tough times and an invitation to shape our lives around shared values of loyalty and resilience.

Ultimately, the tribe we leave behind is a cornerstone of the book's central premise: The book's central premise explores how combat addiction leads humans to seek out combat again. War bonds combine with the addictive nature of battle as they embody the search for belonging and trust rather than violent acts. As we move forward, we must embrace the answer: We must discover ways to respect tribal wisdom while redirecting our deep-seated need for belonging toward a meaningful and peaceful existence. The tribe's legacy exists within us today because we preserve it as an ongoing part of who we are.

Chapter 6: Calm in the Chaos

On the surface, War and conflict would seem to embody chaos, a breakdown of order and peace. And yet for so many people, particularly those who have endured the furnace of combat, there is a surprising reality: profound calm amidst the chaos. It is no accident—it is strongly associated with humans' fundamental nature, which, over millennia has evolved to look for meaning and simplicity during crisis.

Conflict nurtures the primitive human desire to defend and protect. Conflict heightens our focus, reducing the messiness of life to a single objective: to live. Amidst the turmoil of war, the mind eliminates distractions, focusing on what is important. Heart rhythms race, adrenaline courses through the veins, and the senses sharpen to almost supernatural levels. This increased state of focus, forged out of chaos, brings about a sense of calm and mastery in the storm.

Why Humans Crave Conflict

The tension of battle is relentless. Even before a shot is fired or a command is given, the mind is poised and alert, anticipating danger and the duty to react. This constant assessment of risks forms a tightening grip of anxiety. Awareness intensifies, and every nerve is on edge, bracing for action. It manifests as hyper-vigilance, burdening both body and mind, while an inner storm rages invisibly.

When the moment of confrontation finally arrives, a storm of emotions ignites. The clash itself—a fundamental instinct embedded in human DNA—triggers a cascade of physiological and psychological reactions. Adrenaline surges through the body, enhancing focus and numbing pain. The brain produces endorphins, creating a feeling of euphoria in the midst of peril. Testosterone surges through the body increasing strength, stamina, and mental resilience. Time appears to stand still, transforming the chaos of battle into moments of breathtaking clarity. What felt overwhelming just moments ago now seems attainable, almost instinctual.

This release of tension is not only physical; it is also psychological. Battle dispenses with the distractions of everyday existence and leaves us with what matters most: survival, camaraderie, and mission. At these moments, warfighters will describe experiencing a profound sense of peace, as though the chaos around them has been muffled. The certainty they experience is something they cannot find in ordinary life, a period where instincts drive them, and the mind is free from doubt and hesitation.

The sensation is transformative. Amid the maelstrom of battle, warfighters discover a sort of purity, connecting them to their most primal selves. Action is purposeful, every choice matters. There is no time for second-guessing or thinking twice. There is only the immediate, raw, visceral nature of things. The purist experience is one of intoxication, and for some, it is the standard by which everything else is compared. The most comparable description that could be made is "orgasmic" which is described as the rapid, pleasurable release of neuromuscular tensions at the height of arousal. Hence the title of this book, 'Wargasm'.

But this peacefulness is more than just surviving. It is also based in the bond established in combat. In combat, soldiers rely upon one another on a level greater than words can convey. Whatever is built during this period by trust, honor, and companionship, produces an unbreakable bond that defines the experience of being concentrated and sharp on intention. It's a tribal connection that appeals to the deepest aspects of human nature, reminding warfighters they are part of something greater than themselves.

For most, however, the longing for that tranquility endures long after the war. Combat's simplicity and relief are jarring compared with the chaos of life outside the battlefield, in which the strains of bills, routines, and daily pressures both overwhelm and render useless. Combatants describe routinely a wistful yearning for the combat zone—not the carnage, but the camaraderie of shared purpose and membership they felt during the chaos.

This hunger for war is not glorification of war but a picking-up on a basic human truth: that in the chaos of war is a freedom, a feeling, and a peace that is a primal and timeless thing. It is this paradox- the calm within the storm- that lies at the center of why men are drawn to war and why warfighters so often find themselves missing the cut-and-dry nature of those times after the echoes of the battle have died away.

Attempts at Suppressing the Instinct

Throughout history, humanity has attempted to suppress the desire for war, reaching for peace as an objective—a way of avoiding the destruction and chaos that ensue because of war. Regrettably, history portrays peace as brief, its potential unmet by the forces of the very tendencies it seeks to suppress. The tension between the desire for peace and the natural human tendency to fight is central to this dilemma, making long-lasting peace impossible.

One of the earliest mechanisms through which societies moderated the insanity of war was to establish truces and rituals. In Greece, the Olympic Games were not just a festival of athletic prowess, but also an occasion when city-states in conflict declared a temporary peace. These practical truces offered brief periods of peace that allowed communities to be reunited and recover. But they were pauses, not solutions, for the underlying tensions and hostilities had still not been addressed.

Religious doctrines from across the globe have also tried to guide humanity towards peace. The Buddhist philosophy of ahimsa, or non-violence, is the invitation to renounce harm in thought, word, and deed. Christianity instructs forgiveness and turning the other cheek. Islam, in its authentic expression, instructs peace and surrender to God's will. These religious systems have provided ethical direction, calling on higher principles to override basic tendencies toward violence. But even within religious societies, war has repeatedly broken out, with disagreements over interpretation, power, and conflicting ideologies rekindling the fire of discord.

Modern peacebuilding initiatives, from pacts and coalitions to international organizations like the United Nations, represent human activities' broadest attempts to overcome the war drive. Peace initiatives such as the Treaty of Versailles, the League of Nations, and later the UN have sought to prevent conflict by fostering negotiation, accountability, and collective security. These initiatives have secured temporary spates of peace but rarely delivered the extended peace they had hoped for.

The reasons for such failure are structural as well as psychological. Structurally, peace agreements fail to address the embedded frustration and imbalance that fuel conflict, instead treating symptoms and not causes. Psychologically, the human mind cannot tolerate the stillness of peace. With the lack of clarity and concentration born of turmoil, men and societies are apt to be beset by tedium, fidgetiness, and a need for the thrill of challenge. As Sigmund Freud observed, the instinct of aggression, although suppressed, is never absent, manifesting itself in novel forms when war is not present.

The psychology of peace strikes a particular chord with warfighters. Returning from the battlefield, many are jarred by the stillness of civilian life. The mundane details of daily life—paying bills, navigating social etiquette, planning for the future—can be overwhelming and irrelevant in contrast to the stark intensity of war. The absence of shared struggle and war's camaraderie leaves one feeling hollowed out, and they miss the meaningful chaos they left behind.

All through history, this paradox has shown up as wars and peace in a cycle. The 'Pax Romana,' a long time of peace in the Roman Empire, was not a product of war renunciation, but due to the ascendancy of an oppressive power over others. Similarly, the Cold War had a clumsy peace, sustained by threats of destruction by both sides over people. These examples show that peace, while desired, often relies on the careful management of conflict rather than its eradication.

The weakness of peace is that it cannot meet the human need for conflict, focus, and solution. With no guarantee of chaos, peace can seem to be stagnation—a vacuum that sooner or later will have to be filled. This paradox renders efforts towards lasting peace bound to fail since human nature grapples with its two-natured essence: peace and war.

Modern Manifestations of Conflict

In the world today, the primitive war urge has not been eliminated; it has simply evolved. Without war as a steady part of most people's lives, humans have found alternative means to channel this energy: competitive sports, business rivalry, political parties, and even forms of entertainment. These modern-day tribal outlets offer people all parts of conflict, giving them meaning and identity without the physical risk of war. However, they often fail to live up to bringing that same deeper peace and sense of clarity into their lives amidst the war's chaos.

In combat, everything is on the line immediately and primally. The objective and purpose of a warfighter is distilled down into one driving purpose: staying alive and keeping others alive. The adrenaline coursing through their blood sharpens their senses, and the trust in their fellow comrades creates an indomitable bond. Time is frozen, and every decision is laden with consequences. This increased level of perception takes away the noise of life's distractions and leaves only profound clarity and tranquility behind. The chaos around them is contrasted with an inner peace, where there is only the present moment.

On the other hand, modern forms of conflict are rarely so clear-cut. Sport is able to mimic the tribalism of war, with teams and fans uniting under a shared flag. Adrenaline rushes run through the players, and victory brings fleeting moments of bliss. But rarely are the stakes as basic as life or death. While emotionally compelling, they lack the existential weight that characterizes combat, and the peace and feeling of clarity in warfighters' lives is still out of reach.

The corporate battlefield also offers another platform where competition and strategy in warfare are mimicked. Business leaders and their companies compete for market dominance, surmounting challenges with strategy and determination. But the chaos of the business world is also inextricably blended with bureaucratic politics and obfuscate goals. Competition is confused by competing profit margin demands, brand control, and strategic planning. Unlike war, in which immediate action and stakes are concrete, corporate combat is murky, with victories and defeats obscured by an undertow of abstraction.

For the average professional, the pressures of modern life overlook the sense of peace and euphoria that fighting brings. Even though the rush of competition gives us flashes of excitement, it is not so untainted and simple as in fighting, where every movement relates directly to staying alive and tribal unity.

The Division of Modern Politics and Its Chaotic Noise

Political propaganda also taps into tribal instincts, with individuals aligning themselves with parties and ideologies that reflect their values. Contemporary political discourse serves as an example of conflict theory as it provides a setting for ideological confrontation through debates and electoral competition. Political competition creates an environment in which individuals knowingly engage in confrontational behavior rooted in the adversarial approach of their political systems. Those engaging in political debates experience emotional arousal; likewise, the significant emotional responses in combat are comparable to those derived from the relationships scrutinized in this context. Intense debate about candidates and parties provides an opportunity for a similar physiological response in the political context as seen in the warzone. Election competition exacerbates the practices of this theory as candidates engage in contentious campaigning for the opportunity to gain electoral success. The political sphere mirrors battlefield practices in the struggle for influence and governance through these constructs. It exposes the concept of conflict theory in its most long-standing and enduring settings: a society seeking control. Yet, in modern politics, the battle is often fragmented. Social media amplifies voice, and there is a cacophony of opposing views and an echo chamber. This infinite loop of argument and division rarely results in resolution, and the bonds formed in political tribes are less intense and immediate than those formed in combat.

While political activism can be a source of meaning, it almost never provides warriors' peace during war. Political stakes in conflict are high, but the interminable dance of cross-purposes and the lack of final outcomes leave players spent and disillusioned. Political struggles don't often prove as decisive as the battlefield, where a turn of a moment grants victory or loss.

Entertainment: A Safe but Hollow Reflection

The most abundant modern expression of conflict may be in entertainment. Films, video games, and books transport audiences to fantasy battlefields, where they can experience vicariously the thrill and triumph of warfare. While these forms of media give expression to humanity's innate urge for conflict, they are simulations, which are safe, controlled settings that cannot reproduce the full gamut of feelings experienced in real war. Ultimately, the virtual combatant knows there is no real danger and thus has no real opportunity to build and release the same level of tension as actual combat. Action movies usually project stories about heroism and survival, and allow moviegoers to experience faux combat situations, without the actual danger of being harmed. Video games go a step further as they allow the players to be participants in combat situations and experience the same thrill of utilizing those instincts in a desire to win. In this sense, the same human instincts have been accommodated into these forms of entertainment, thereby exhibiting society's innate human instinct to engage in actions that fire their excitement.

For others, such estrangement from reality is satisfying and comforting. The simplicity and peace warfighters feel in actual combat derive from the gravity of their circumstances, immediate access to their instincts, and reliance on their comrades. Whatever the immersion or thrill, entertainment cannot compete with the depth of these experiences, with participants craving something more real and profound.

Conclusion: The Paradox of Calm in the Chaos

'Calm in the Chaos' is the name of the paradox at the heart of the human condition of war. Chaos appears destructive, but it produces times of clarity, purpose, and unity that demystify the simplicity of the complexity of life. Whether in war, competition, or the everyday turmoil of modern times, humans retreat to these moments of heightened existence. This irony speaks to the ongoing struggle between war and peace, reminding us that chaos and serenity are not opposing forces, but complementary opposites that define human existence.

While modern forms of violence are full of purpose and tribal belonging, they lack war's unambiguous simplicity and rapture.

These spaces are typically splintered, caught up in complication, or severed from the primal plane of stakes that defines war. For warriors, the peace they find in war is directly proportionate to the singleness of its anarchy: survival, camaraderie, and raw exposure to life and death. With all its strife and tensions, modern life can't seem to replicate this balance, so many still long for the certainty they enjoyed in the melting pot of war.

Chapter 7: Lessons from the Battlefield

The experiences on the battlefield offer a unique lens into the behavior of individuals when placed in extreme situations, thereby challenging and redefining our perceptions of innate human traits. According to Rutger Bregman, war prompts a reassessment of "our basic assumptions about human nature," suggesting both adverse and favorable characteristics are revealed in times of conflict. Moreover, the societal upheaval caused by war provides insights into how communities respond under pressure, often resulting in increased social cohesion and religiosity. By evaluating these aspects, we can recognize that war serves as a catalyst for both individual transformation and collective adaptation, leaving indelible marks on humanity and compelling us to confront both the good and the bad of human existence.

War undeniably stands as a crucible for examining the multifaceted impacts on both individuals and societies. As individuals face the extremities of conflict, war becomes a profound teacher, illuminating the strengths and vulnerabilities inherent within the human psyche. Societies, similarly, are placed under immense pressure, revealing latent capacities for unity and resilience, juxtaposing both cooperation and division. These reflections on human nature garnered from war prompt significant contemplation about societal values and priorities, as communities adapt to shifting realities brought forth by conflict. Thus, the lessons learned from war extend beyond the immediate, encouraging a deeper consideration of how humanity might forge a path toward peace while grappling with the undeniable effects of its engagement in warfare.

In examining what war teaches about human nature, one must consider how extreme conditions on the battlefield offer profound insights into human behavior. War exposes both the virtues and flaws inherent in individuals, revealing tendencies that are often cloaked in peacetime. War challenges societal functioning, as communities are forced to navigate and adapt to unprecedented pressures. The outcomes of such collective trials underscore the resilience and adaptability of societies in the face of adversity, highlighting humanity's extraordinary capacity to evolve and reorganize amidst turmoil.

The pressures exerted on societies during war often illuminate the underlying dynamics of communal responses and adaptation. Conflict serves as a critical period when traditional societal structures are tested and reconfigured, thereby revealing both strengths and vulnerabilities. Societies frequently exhibit an increased capacity for unity and resilience, as war necessitates collective action and coping mechanisms, often leading to enhanced social cohesion and religiosity. Conversely, the strain of warfare can exacerbate existing divisions, prompting a re-evaluation of social values and priorities. Through these dual processes, war acts as a lens through which ways that societies withstand and adapt to adversity are scrutinized, providing essential insights into the broader human condition.

The internal burdens borne by warfighters illustrate the profound personal and psychological costs of war. The harsh realities of conflict inflict lasting emotional and mental scars, reshaping the very essence of the individuals who endure such experiences. Bregman emphasizes the duality of human traits manifesting in extreme conditions, highlighting how war tests not only physical endurance, but also psychological resilience.

As soldiers navigate the chaotic milieu of the battlefield, they encounter moral dilemmas and witness unspeakable horrors, which can lead to conditions such as post-traumatic stress (PTS). This psychological toll underscores the need for comprehensive support systems aimed at facilitating recovery and reintegration, as the psychological impacts of warfare extend well beyond the field of combat, influencing real-life social interactions and future societal contributions.

The devastating impact of war extends far beyond individual experiences, profoundly altering entire communities and global societies. Human suffering during warfare becomes a powerful testament to the broader implications of conflict, prompting a reassessment of human resilience and vulnerability. Wars often leave a traumatic imprint not only on the immediate victims but also on future generations, reshaping collective identities and influencing societal structures for years to come. As societies reorganize and respond to the devastation, they encounter opportunities to fortify values and priorities that may lead to pathways of recovery and peace—a crucial consideration in understanding how humanity might evolve post-conflict, ensuring that the lessons learned contribute to a more empathetic global society.

What War Teaches Us About Ourselves

Revealing Strengths and Vulnerabilities

War acts as an unyielding mirror, compelling people to face their genuine selves. Combat situations make every action seem more intense and every choice permanent. Many individuals discover previously hidden sources of strength and bravery during these critical moments. Warfighters push past their perceived boundaries and uncover resilience, which necessity brings to light.

But combat also exposes vulnerabilities. Combat removes soldiers' illusions of invulnerability and forces them to face their fears and doubts while dealing with the consequences of their actions. These experiences often force warfighters to reconcile two versions of themselves: Warfighters face the contrast between who they were when they entered combat and who they became after it ended. War demonstrates its transformative ability through the duality of self-redefinition, presenting both empowering and challenging outcomes.

Moral Clarity and Complexity

War illuminates moral issues by pushing individuals to make decisions that test their ethical principles. During combat situations, ethical boundaries become indistinct and survival-driven actions create enduring moral impacts on individuals. In combat situations, some individuals achieve moral clarity because protecting their fellow soldiers becomes their highest priority. The end of active combat does not erase the moral dilemmas some soldiers face, which remain present in their lives.

Ultimately, the battlefield offers a stark truth: The human experience encompasses deep empathy as well as extreme aggression. Soldiers' decisions during battle, followed by their personal reflections, expose the complex equilibrium between compassion and violence inherent to every person.

What War Teaches Us About Society

Unity and Division in Crisis

War possesses a remarkable power that brings societies together like no other force. In response to an outside danger, communities unite to put aside their differences and work towards a common goal. In these moments, war can reveal the best in humanity: acts of heroism, self-sacrifice, and collective resolve. The collaboration and shared purpose shown during wartime demonstrate how unity can emerge from challenging situations. In the aftermath of the terrorist attacks on September 11th, 2001, America was united in a way that had not been seen in generations. Echoes of "United we stand" chants could be heard from coast-to-coast.

War reveals existing social divisions and inequalities. People who face marginalization or vulnerability experience the greatest challenges from higher rates of frontline military service to devastation within their communities. Societal divisions which remain hidden during peacetime become evident through leadership failures and corrupt practices.

The Fragility of Peace

Through its lessons, war demonstrates to society how brittle peace can be. Humanity has made progress in diplomacy and governance, yet conflicts persist because of unresolved grievances, competition for resources, and inherent dominance drives. War pushes societies to analyze their systems and principles through difficult questions about conflict origins and future prevention strategies.

Societies that emerge from war often face a crossroads. After conflict, societies have the option to create sustainable peace or revert to patterns of violence and division. Post-conflict choices demonstrate whether that society's social fabric shows resilience or fragility.

Warfighters Face Considerable Internal Costs After Combat

The Weight of Trauma

Learning from the battlefield experience demands a significant personal sacrifice. The visible injuries sustained by warfighters during combat often pale in comparison to their hidden psychological wounds. Post-Traumatic Stress (PTS) frequently manifests through recurring flashbacks, persistent anxiety, and increased vigilance. The mind battles to align war's turmoil with civilian life's tranquility through these symptoms.

Warfighters who endure PTS can also suffer from moral injury which manifests as guilt or shame because of their actions or experiences in combat. The internal struggle over war experiences creates a sense of isolation among individuals as they confront their wartime actions and their roles in combat situations.

Longing for Clarity Amidst Chaos

Many veterans describe combat as a paradoxical experience: Combat presents a frightening and chaotic environment which simultaneously delivers a level of purposeful clarity that civilian life fails to match. The intense bonds with comrades and survival focus combined with moment intensity delivers deep fulfillment in meaning despite ongoing destruction.

Many military personnel find it challenging to regain the purpose they experienced while serving after they transition back to civilian life. The everyday tasks of paying bills and planning ahead while maintaining social norms often appear small and burdensome against the background of war experiences. The gap between wartime experiences and civilian life causes many veterans to miss the clear purpose they found amid war chaos while they attempt to recover from its emotional damage.

The Cost of War on Humanity

The Toll on Lives and Communities

War creates its most direct and observable damage through its effects on human lives and community structures. The destructive legacy of death and displacement alongside ruined landscapes results in wounds that require multiple generations to recover. War destroys cities and cultures while separating families. Survivors endure an immeasurable physical and emotional toll as they work to rebuild their lives while simultaneously coping with their grief and loss.

The destruction caused by war extends beyond its initial impact to create enduring economic and political challenges. Rebuilding infrastructure, revitalizing economies, and reestablishing governments represent a multi-decade challenge that frequently encounters major obstacles. The ongoing existence of landmines, along with environmental destruction and other conflict remnants, demonstrates how war creates an enduring shadow.

The Psychological Ripple Effects

The psychological trauma caused by war affects not just direct participants but reaches society at large. Collective trauma burdens entire societies when conflict narratives become a part of their cultural memory. The historical impact of war on art and literature affects public awareness and guides the way future generations understand war and peace.

The pattern of repeated violence tends to sustain its existence. The unaddressed animosities and societal splits following war become the breeding grounds for upcoming conflicts. The repetitive cycle of war requires humanity to develop new approaches that resolve tensions before they lead to conflict.

Lessons for the Future

The destructive impact of war juxtaposed with its clear outcomes leaves humanity with two distinct legacies. The text demonstrates how human nature can reach destructive extremes. Concurrently it demonstrates how resilience functions alongside adaptability to create transformative opportunities. Understanding this paradox helps us dismantle violent patterns and create a future where conflicts lead to positive outcomes instead of destruction.

Redefining Conflict and Purpose

The essence of war demonstrates that human beings possess an inherent inclination towards struggle and competition while showing that the results of these tendencies remain uncertain. Historically, the demands of war have led to significant innovations including medical advancements like penicillin and groundbreaking technologies such as the internet. The lesson is clear: The natural human tendencies toward battle and ambition can be transformed into productive force for creation and advancement. Creating spaces where challenges and competition fuel progress across scientific research, educational systems, and artistic endeavors, enables us to transform fundamental human drives into positive forces for humanity.

During times of peace, the primary difficulty revolves around helping people discover purpose as war naturally provides. By focusing on shared objectives through environmental conservation initiatives and collaborative enterprises, institutions can recreate the connection and urgency that war produces. Humanity's inherent desire for struggle finds purpose when programs address global challenges like climate change and resource scarcity while promoting collective progress and unity.

Breaking Cycles of Violence

The constant recurrence of war due to the persistence of unresolved conflicts stands as a major obstacle for human civilization. Societies can dismantle these harmful cycles by putting reconciliation, equity, and opportunity at the forefront of their priorities. Truth and reconciliation commissions provide vital structures to address past wrongs while promoting mutual understanding among divided groups and averting future tensions. These processes demand immense effort, but they establish the essential groundwork needed for durable peace.

Policies focused on economics and society need to target inequality along with systemic barriers because these elements frequently serve as conflict origins. Societies will reduce the desperation and resentment that cause unrest when they invest in equitable education alongside infrastructure development and sustainable growth. When marginalized communities gain power, it creates more inclusive and stable societies that prevent war from exploiting people's grievances.

Harnessing Collective Strength for Global Challenges

The principles of discipline, adaptability, and unity that emerge from wartime experiences provide a strategic guide for tackling worldwide problems. Historical wartime collaborations show that international cooperation becomes essential during global crises like pandemics as well as environmental and resource challenges. The Paris Climate Agreement shows how nations must unite to address climate change but trust and accountability between countries still need significant improvement.

The future calls for a shift in mindset from group competition toward species-wide collaboration. When we recognize the potential for unity within our common struggles, we direct human survival instincts and ambitions toward establishing a cooperative world instead of a conflict-driven one.

Conclusion

The battlefield embodies deep contradictions by presenting instances of clear insight alongside limitless destruction. Conflict reveals our strengths and weaknesses while demonstrating society's ability to endure hardship and showing the severe impact war has on people and humanity altogether. Reflection on these lessons teaches us to integrate our natural instincts for conflict with our peace goals to create a world that respects historical sacrifices while working toward a compassionate tomorrow.

Chapter 8: Embrace the Suck

The phrase "embrace the suck" reflects a truth every warfighter knows: While hardship is inevitable throughout life, it retains the potential to produce meaningful outcomes. Combat provides uncompromising insights into survival principles while enhancing mental clarity and purposefulness. The real challenge emerges after combat ends when enduring instincts persist and seek meaningful expression. The process entails recognizing difficult situations by facing discomfort head-on, which leads to personal growth through pain. Warfighters transitioning from combat find in this framework a method to transform chaos into clarity while turning hardships into strengths and instincts into meaningful purposes. This chapter addresses transforming war's primal instincts and deep experiences into meaningful paths which support warfighters and society in building purpose after war. Warfighters generally find the concept of "embrace the suck" essential in managing their difficult shift from military service to civilian life. By developing this mindset through exposure to difficult settings, people can turn challenges into chances for self-improvement and creative thinking. Those who embrace challenges during their transition can redefine their identities to achieve meaningful roles outside of military life. This transformative approach pushes veterans to use their past experiences as a basis for future innovation, which builds their resilience and creativity while they establish new civilian life paths.

Understanding the Instinctual Pull of Conflict

The Warfighter Mindset: A Return to Primal Origins

Combat strips life to its raw essentials: survival, the tribe, and the mission. Warfighters discover heightened awareness in battle conditions that civilians seldom achieve through their deep bond with primal instincts. The mind achieves intense concentration while relationships with team members become unbreakable, and time appears to become more vibrant with each passing moment. Many people experience pure self-discovery during these moments.

My own 'awakening' stands vivid in my memory. During chaotic times when the world seemed to collapse, I became laser-focused on necessary actions and responsibilities. My primal instincts served a dual purpose. They protected me and ensured the survival of others. The shared purpose from my past experiences continues to be a part of who I am today. The combination of innate instincts with defined purpose creates a feeling of home for combatants despite the battlefield setting. The clarity we seek draws us in, not the destruction around us. The single-minded concentration turns chaotic situations into a tranquil and euphoric experience. The battlefield leaves such a deep mark on many people that they feel an overwhelming urge to return.

The Instinct to Return to the Fight

Even after missions finish their course, the ingrained instincts stay while the need for clarity persists as an unrelenting ache. I've felt this pull myself. Once back from the battlefield, the world felt overwhelmingly noisy and crowded.

The constant background noise of civilian routines contained complexities that did not possess the pressing nature or meaningful purpose I had become familiar with. The daily responsibilities, along with casual conversations and weekend activities, paled when matched against the camaraderie and mission of combat.

The opportunity to return to combat presented itself in 2008 after my two-year absence from military operations. The two years of civilian life presented its own challenges as I faced daily routines without military structure and intensity, which had been the foundation of my previous existence. At that time, I was unaware of the fact that I had been existing in a sort of limbo since I carried my acquired combat skills and instincts, but did not understand how to use them in my new civilian environment. The term "fish out of water" can't truly describe the feeling. I felt like a man torn apart, separated from my soul and what I felt defined me as a human. For my entire adult life, I had known nothing but war. Being in a world that gave empty thanks, but no real opportunity was a contradiction of my spirit. Stepping back onto the battlefield gave me a sense of starting anew. Reconnecting with my purpose and strong group bonds alongside intense focus brought me great relief.

That final deployment provided me with benefits that surpassed simple relief and perspective. I started to understand that the combat skills I acquired could help me not only survive but also excel in life. When the moment came to reintegrate myself into civilian life I approached it with a transformed perspective. The knowledge I gained from my experiences shaped my actions. Every decision made in my journey forward became a chance to apply and demonstrate the resilience and leadership skills I developed during wartime chaos. The reintegration process now required carrying battlefield lessons with me to create something new. Only after some time did I recognize this pull was not a flaw, but a testament to my strong connection with my instincts. Life changed direction when I decided to embrace the pull rather than resist it. The clarity I sought emerged in different forms through building meaningful connections, new challenges, and assisting others in discovering their own purpose.

The knowledge I gained became a fundamental element that shaped my coaching career. The fundamental instincts that cause individuals to return to conflict situations because of their need for purpose and intense connection can instead be used to create constructive ventures. While supporting others to rediscover their purpose I experienced a familiar mission-focused mindset from my military days which brought me renewed clarity about my direction.

Embracing the Suck and Finding New Purpose

Acknowledging the Suck

Warfighters face no greater challenge than readjusting to civilian life after military service. Without mission orientation and unit support, one experiences the terrifying sensation of plunging through the air without any safety harness. The early stages of my transition surprised me with unexpected difficulties. The end of my military service brought me not only grief over leaving the military, but also deep sorrow for losing that battlefield identity and the clear purpose combat provided.

"Embrace the suck" became my mantra. The key was not to minimize challenges or pretend the transition was easy, but to accept discomfort and use it as a means to develop. I started asking myself the same questions I used to ask in the chaos of combat: What needs to be done next? How can I lead through this? How do I identify my mission at this precise point in time?

That mindset became my lifeline. I chose to embrace the struggle as a chance to build my strength. I realized that the same survival instincts from combat—resilience, adaptability, and focus—could also help me survive a transition if I chose to apply them.

Redefining the Mission

Purpose is the warfighter's compass. Combat requires that all actions align with mission objectives while decisions reflect responsibility toward both yourself and your team members. The disappearance of that structure leads to a sensation of the world becoming unanchored. What I discovered was that missions continue to exist through redefinition rather than termination. Veterans transitioning into civilian life frequently search for new missions that reflect their fundamental principles and enable them to apply their well-developed abilities. Just like any other addiction, finding a new purpose is a critical component to overcoming the addiction of combat.

For me, the new mission became clear: My new purpose involved ensuring that those who still served had the best training possible to be able to fully do their job and fight the wars of the future. Beyond that purpose, I found meaning in assisting others in discovering their direction through career coaching and network building along with developing platforms for veterans to share their experiences. The military discipline and focus that I developed helped me build a career where I could make significant contributions.

Evolving the mission means building upon its past foundations rather than discarding them. Their mission could involve launching their own business or supporting a cause they feel strongly about. Some people apply their military leadership skills to guide their families using the same dedication they showed their combat teams. No matter what you choose to pursue, the main goal should be connecting your actions with a purpose beyond yourself.

Building a Productive Life with Instincts Intact

Leveraging Tribal Instincts

The bond between soldiers is unique and unmatchable, but the tribal connections from military service should transition beyond military life when the uniform is removed. To regain that feeling of connection, I developed networks with professionals who shared my views. Each LinkedIn post I created and every discussion with another veteran became my method of reconstructing the essential tribal network that shaped my identity.

Through service organizations, mentorship programs, and community projects, people can connect with their instinctual need for tribal belonging. Veterans excel when they become part of larger teams or communities that work toward common goals or solve collective problems. Civilian life presents different types of bonds yet veterans still retain their fundamental need for connection.

Channeling Aggression and Focus

When combat intensity and drive for action are directed constructively, they become a strong force for personal growth. Endurance races and martial arts alongside daily fitness routines serve as outlets for the battlefield energy that once ensured survival. Physical activity established itself as a fundamental part of my routine because it benefits both my physical well-being and mental health. The practice brought back memories of combat discipline while providing me with a concrete method to sustain attention.

In my professional role, I discovered methods to direct that intensity towards my work as a coach. I used precision and determination from my military background to help clients solve complex problems and navigate career changes. Helping others was important, but staying true to my fundamental instincts played an equally vital role.

The Way Forward Begins with Action

Warfighters can maintain their core instincts for clarity, connection, and purpose outside military combat. Warfighters who learn to redirect their essential instincts can construct civilian lives that match their military service in intensity and impact.

The call to action is simple: Welcome to hardship with courage but maintain your identity beyond it. Apply your wartime knowledge to guide others through mentorship while building communities and seeking personal challenges that stretch your abilities. While the battlefield can leave lasting impressions on us, these experiences shouldn't become our limitations.

Conclusion

The saying "Embrace the suck" urges people to face difficulties with bravery together with clear purpose. Warfighters learn that survival skills developed in combat act as essential building blocks for personal growth and leadership while creating meaningful impact. Every warfighter possesses the capability to build a meaningful legacy that celebrates their past by transforming their natural instincts into a purposeful life path.

Chapter 9: Reclaiming the Warrior Spirit

The warrior spirit endures within every warfighter as an unquenchable fire that withstands both time and circumstances. It persists beyond the silence of battlefields and continues through the days of peace. The warrior spirit remains active as an endless reminder of the bonds and teachings produced through intense battle experiences. Reclaiming the warrior means honoring one's past while understanding that wartime clarity and purpose can guide a meaningful and impactful life. The process of transformation involves using experiences from war to create something more significant.

Combat has a way of sharpening identity. Every move taken on the battlefield counts while decisions hold tremendous importance, and teamwork develops through mutual goals and confidence. Warfighters discover their true selves as they come to terms with their endurance capabilities within that transformative environment. When military structure ends with a mission, numerous veterans experience disconnection while searching for clarity and belonging in a civilian world that feels distant from their previous intense experiences. Reclaiming the warrior means rediscovering clarity by creating a new direction from the battlefield ethos and learning lessons instead of recreating military conflicts.

This journey is not without its challenges. Moving from military service to civilian life creates a sense of loss which includes missing the close-knit community of the tribe and the adrenaline-driven mission that fueled their daily existence. But within that loss lies an opportunity: This invitation challenges warriors to transform their identity into an active force for the future rather than clinging to past definitions. Embracing combat-honed values of discipline, honor, and loyalty allows warriors to integrate these principles into all life aspects, including personal relationships, professional ambitions, and community leadership roles.

This chapter examines the process of starting that journey. Our understanding of how to adapt the warrior spirit to modern battlefields arises through examining legacy and ethos while learning from mentors and engaging in spiritual reflection. The passage serves as a call to action and emphasizes that the distinct clarity experienced in combat can be preserved to become a lasting, powerful legacy. The warrior's redemption lies in discovering a new mission that respects historical roots yet creates a superior future through growth-focused leadership and purpose-driven objectives.

Embracing a Legacy Beyond Combat

In combat, purpose is rarely in question. Each action and spoken word carries essential meaning. Life becomes a sequence of urgent choices where each choice stands out as decisive and carries significant weight. Warriors develop their sense of purpose beyond instinctual responses as this framework helps them discover their identity where actions are linked to survival. Exiting the military environment means not only returning to peace, but also abandoning the most authentic sense of identity we've ever experienced. Over time true warriors maintain their sense of clarity. A warrior rebuilds his legacy by applying battle lessons of discipline and focus to forge a new path through civilian life.

The quiet aftermath of battle resonates louder for many of us, more so than the combat itself ever did. Orders and directives have disappeared, along with unit camaraderie and unified mission moments. The world that follows combat presents a cacophony of civilian distractions that appear absurdly insignificant after experiencing life at its most delicate state. The routine demands of life seem empty when compared to the meaningfulness of war. The loss of clarity and purpose creates frustration that transforms into disorientation.

During my initial months after returning from deployment, I looked at my calendar and questioned how an activity so basic had become so overwhelmingly challenging. In combat, time had structure. The unit's mission and its requirements determined the schedule instead of meetings or time management. Out here, every moment felt scattered, disconnected. The lack of purpose impacted me more significantly than carrying my equipment ever did. The emptiness inside me was unfillable through nostalgia for the past or attempts to recreate combat experiences because it demanded something entirely new and future focused. Through this realization I discovered the concept of legacy.

A warrior's legacy emerges from their post-war actions instead of being defined only by their combat achievements. The battlefield's experiences create lasting bonds that persist beyond conflict as lessons and foundational tools for building future lives. Embracing this legacy requires us to see war-born values like discipline, courage, and loyalty as valuable gifts we can share with the world, instead of outdated relics. Our combat experiences provide the essential base for creating meaningful lives because these experiences themselves give depth to our existence. We were able to tap into the most primal instincts of humanity. Why would we stop using that instinct to continue to be a part of humanity?

The lasting benefits of combat include the powerful bonds of unity and loyalty. The endless nights I spent talking with my brothers led to deep reflections on life's absurdities and profound truths. The hours we spent together were more than idle time because they created a powerful and permanent bond among us. I learned that true strength comes from connections and the assurance that others will protect you just as you protect them. The understanding I gained stayed with me even after my military service concluded. Memories of connection and loyalty provided me with direction when I struggled to find purpose after leaving the battlefield. I looked for methods to recreate that unity by establishing similar bonds with others in my civilian life instead of returning to combat.

Mentoring younger veterans became one of the most impactful means through which I sustained my legacy. Their confusion and displacement mirrored my own past experiences, which led me to see a chance to direct them on their individual journeys. Through mentoring I succeeded in transferring past lessons into practical methods that motivate others. My goal was not to reenact combat experiences, but to build a fresh form of camaraderie through shared missions of adaptation and upholding our common values without military gear or weaponry while maintaining the same significance.

Adopting a legacy that transcends combat does not require warriors to abandon their spirit, but instead calls for broadening its meaning. The attributes of strength and courage that define the warrior are equally essential in peaceful times as they were during wartime. Choosing consciously to use combat lessons as foundational elements for personal growth and the development of others demonstrates this principle.

Warriors learned to stand together as a unit to withstand difficult situations. Transitioning to civilian life means moving forward with combat-acquired traits to create new opportunities instead of dwelling on past experiences. This is the essence of legacy: Drawing strength from past experiences enables us to concentrate on future possibilities.

To rebuild purpose and identity in this new world, we need to understand the framework that directed us during wartime. The warrior ethos serves as our framework and code, establishing both our actions and our identity. The path to reclaiming the warrior requires us to implement that warrior ethos in our current world.

The Warrior Ethos in a Civilian World

The warrior ethos extends beyond the battlefield rules because it represents an intrinsic part of our identity that directed our actions during critical combat situations. This code stands upon core principles of discipline, honor, and loyalty together with a steadfast dedication to higher ideals. In war, the ethos is non-negotiable. The warrior ethos acts as the guiding force behind all movements and decisions while shaping every thought. Outside military service, people confront a world that rarely asks for the same clarity or high stakes found in battle, yet they face the challenge of applying their battle-honed values to a civilian world that seems to have little need for warriors.

But the warrior ethos is not situational. It exists independently from any battlefield presence. The warrior ethos establishes itself as a mindset that guides life choices while remaining a pledge to adhere to timeless values regardless of life's direction and stays relevant even when conflicts end, and weapons are no longer used. The significance of this ethos becomes even more vital in a world characterized by widespread detachment and complacency. The process of reintegrating the warrior ethos into daily civilian life requires acknowledging its eternal relevance while utilizing its principles across all aspects of life regardless of their distance from martial settings.

The concept of honor becomes unfamiliar when people leave their military service for civilian life. The warfighter's sense of honor stands out as a refreshing change in today's world where individual interests typically overshadow shared duties. Honor represents more than integrity because it requires responsibility towards something greater than oneself. When facing combat situations, the focus shifts to completing the mission goals or keeping faith with the team and protecting core values. In civilian life accountability becomes a commitment towards honest leadership and service to others. A warrior's dedication to righteous action regardless of costs generates waves of trust and respect across all environments. I discovered that honor integrated into my civilian life meant more than big achievements because true leadership came through daily decisions. I needed to fulfill my promises while owning up to my errors and maintaining my core values regardless of others' awareness.

Within the warrior ethos, loyalty stands out as the most important factor that connects us to each other and our mission. The battlefield transforms loyalty into an essential survival tool rather than just a theoretical concept. The trust you put in your comrades with your life comes full circle because that trust is always returned. It's unspoken but absolute.

As I made the transition to civilian life, I quickly noticed the loss of loyalty, which made conversations feel empty and relationships seem unstable, while trust became something that was not freely given or received. This concept started to reveal new possibilities for reconstruction as I reflected on its true meaning. Instead of recreating combat bonds, this new stage demanded the same dedication to forge meaningful and lasting connections. True loyalty meant supporting people when they faced uncertainty while building friendships that endured over time to establish communities that echoed the battlefield camaraderie I experienced. Loyalty emerged as an enduring principle that you brought into every interaction and decision because it remained with you after war.

Mentorship: Passing the Torch

No warrior fights alone. The strength of your survival in battle relies heavily on your connection with your fellow soldiers who stood with you through the trenches and helped carry your fears and burdens when they became overwhelming. The sacred bond between warriors develops through warfare, and they survive intense stressful situations. What becomes of the bond soldiers share once hostilities cease? The loss of battlefield camaraderie stands as the most painful loss for many who experience it. Mentorship provides a pathway to restore that connection by enabling individuals to transfer battlefield values and teachings into new essential missions. The focus here isn't on remembering the past, but rather on handing off the torch to others so the warrior spirit continues through new lives.

Entering a mentorship position initially left me uncertain about my contributions. How can I express something meaningful to a newcomer who has yet to encounter the challenges I faced? The essence of mentorship resides in its ability to transcend grand gestures and perfect words. It's about showing up. The key aspects of mentorship include active presence and attentive listening while providing consistent support that reflects what you received from others. When you express, "I've been there and I'm here for you," you create transformative moments for both the mentee and yourself. It reinforces how your past experiences taught you valuable lessons that proved worthwhile through the difficulties you faced. They represent fundamental elements that contribute to a more significant and enduring creation.

Mentorship represents an ongoing mission rather than a charitable action. The experienced warfighter on the battlefield must secure their own survival while also directing and shielding their inexperienced teammates. The duty one bears as a warrior persists beyond the removal of military attire. If anything, it grows stronger. Our combat experience gives us valuable knowledge about resilience and decision-making under pressure, along with the importance of unity, which provides others with the tools to face their own difficulties. The dissemination of our lessons allows warrior ethos values to transcend history and drive future developments.

I've seen this power firsthand. I've seen mentorship generate positive impacts through brief conversations that build confidence in people who feel lost, and shared moments of vulnerability, which create deeper trust and connections. I recall guiding a young veteran who was having difficulty adjusting to life after coming home. The young veteran experienced similar feelings of disconnection and aimlessness that affect many individuals. We analyzed his experiences together with the goal of discovering meaningful purpose rather than focusing on past events. The greatest fulfillment in my life came when I saw him discover his own power and understand the importance of his story. I served as a guide to help him discover his innate potential rather than instruct him on what to do.

Mentorship is a reciprocal process. We frequently view mentorship as a one-way exchange where we serve others, yet it actually returns equal value to us. The mentorship experience allows you to rediscover your essential identity as a warrior leader while validating the importance of your past experiences. Combat lessons remain active and essential truths which continue to direct and motivate us. Mentorship allows us to discover a fresh mission which holds equal significance to our military endeavors.

As warriors, we recognize that true strength emerges through connections rather than solitude. Mentorship is a powerful method for restoring connections while fostering a sense of unity and camaraderie in a civilian world that frequently feels divided. The challenge before us requires us to support others by sharing our knowledge while leading future leaders and warriors through mentorship. Mentorship allows us to restore battlefield connections by establishing new missions that embody our core values instead of returning to warfare.

The mentorship model stays consistent across all environments, whether it's military combat or corporate leadership, because warriors always operate in teams. By passing the torch, we pay tribute to our past while guiding others toward their future destinations. The warrior ethos survives through active preservation rather than historical remembrance.

The Spiritual Dimension of the Warrior

Warfare forces individuals to confront life's deepest existential questions. War chaos eliminates daily trivialities as life hangs by a thread and reveals our fundamental human reality. For many warriors, this confrontation brings clarity and insight into their personal boundaries while discovering their purpose in life and accepting their mortality. The experience imprints a spiritual scar which continues to challenge warfighters to interpret their experiences even after the battles conclude. Reclaiming the warrior requires more than honoring discipline and loyalty because it demands finding significance in scars and turning pain into purpose while merging war lessons into a broader spiritual framework.

The fundamental nature of the warrior's journey embodies spiritual elements. Entering a battlefield awakens primal instincts within us while establishing a profound connection to something larger than ourselves. That same intensity leaves its lasting imprint. Combat decisions alongside endured losses and nightly replaying memories leave permanent marks on the soul. Moving back into civilian life causes those deep grooves in our souls to become even more pronounced as we struggle with questions that lack simple resolutions. Why did I survive when others didn't? How can one lead a life that honors and respects the ultimate sacrifice they made? How can I bridge the gap between my former combat identity and my present self?

The core of reclaiming the warrior identity lies within these questions rather than existing as distractions from it. Visible and invisible scars demonstrate our resilience and represent the challenges we have faced and conquered. The spiritual warrior accepts their scars while discovering how to use them to influence their character constructively. We gain insight not through war's brutality, but from the resilience and courage it taught us and our ability to withstand hardship. The lessons of war become guiding principles that direct us through life's complex challenges outside combat zones.

The spiritual journey reaches its deepest impact through the practice of honoring our departed loved ones. Warriors never forget their fallen comrades who never returned home. The faces we remember along with their names and voices become permanent parts of who we are. These memories generate both pain and purpose for many people. We honor the fallen through remembrance and by embodying the principles they lived for. We honor their memory by choosing paths that create positive change while healing wounds and raising others up.

The moment this realization overwhelmed me remains vivid in my memory. I stood next to several of my Marine brothers at the grave of one of our fallen brothers. His loss weighed heavily on me while simultaneously providing me with clear insight. The sacrifices he and the others made served as powerful reminders of the character I aspired to achieve. From that moment on, I began to see my scars not as burdens but as reminders of my responsibility: I have decided to live my life to the fullest while demonstrating integrity as a leader and leveraging my experiences to benefit the world around me.

The spiritual aspect of being a warrior requires maintaining deep ties with the community. During combat situations, we always remain connected because our actions contribute to a bigger mission and our choices influence our team members. The interconnection we felt during war persists through the civilian bonds we form afterward. To reclaim the warrior mindset requires building connections between people and establishing places where we can unite through mutual support and collective goals. We discover that the power we drew from our military comrades continues to exist through the communities we establish with veterans, family members, or strangers united by shared goals.

Embracing the spiritual warrior path requires facing forgiveness towards others as well as self-forgiveness. Many of us carry guilt from our time in combat because we took lives during combat, or made difficult decisions while others died, or that we survived. The weight of our guilt acts like an anchor that pulls us downward while holding us bound to our past experiences. Forgiveness requires us to acknowledge our human nature while creating space for future progress. The path to honoring our past requires us to understand that our mistakes do not determine our identities, and we should focus on leading lives filled with purpose and compassion today.

To reclaim the warrior is to embrace all aspects of who we are. Our combat experiences demand discipline and courage while loyalty and unity connect us to our comrades. Our enduring scars and questions serve as reminders of our past trials. The warrior's spiritual dimension connects all these elements while providing meaning and direction throughout our shift from war to peace. This side of ourselves prevents our experiences from being diminished to mere pain or regret while turning them into wellsprings for strength and wisdom and compassion.

The warrior path continues to develop after battles conclude instead of coming to an end. Although the battlefield becomes distant from us, the teachings it provided us continue to persist. Reclaiming the warrior involves taking lessons into our future lives as guides rather than burdens. The spiritual dimension shows us that our scars reveal our resilience and demonstrate our survival and ability to contribute valuable lessons to others.

Conclusion: A New Mission for the Warrior Spirit

The warrior's path to self-discovery extends well past military combat. The transformation journey begins as a conscious change by which individuals convert wartime values into tools for constructing a significant and lasting existence amid peaceful complexity. Warfighters can discover both clarity and purpose in civilian life through each step of their journey, which includes embracing a legacy beyond combat, living by the warrior ethos, alongside mentorship, and spiritual identity aspects. These elements unite to create a blueprint that blends combat instincts and teachings with strength into a life marked by personal development and influential leadership.

The essential process of reclaiming the warrior spirit represents a spiritual path that enables individuals to reconcile their battle wounds and convert suffering into meaningful objectives. When we adopt the spiritual aspect of the warrior, we link the memories of our lost loved ones and our enduring questions with the burdens of survival and sacrifice into a broader story. The spiritual dimension connects all parts of the journey and provides meaning which moves beyond the limits of war and peace.

The clarity found in warfare need not disappear but can instead undergo a transformation. The skills developed during intense combat situations can continue to grow rather than weaken. The true essence of reclaiming the warrior spirit involves using past experiences to construct future paths. The warrior spirit persists beyond battle through leadership deeds and meaningful service connections.

To reclaim the warrior spirit is to take up a new mission, to build structures while leading others and inspiring their growth. The values warriors learn during warfare stand the test of time, while their combat connections prove indestructible. The warrior spirit has the power to create a more compassionate and enlightened world. The true meaning of reclaiming the warrior means seeing it as a dynamic force that shapes the future rather than just a historical artifact.

References

As with any writing, this work was developed using the work of numerous authors, researchers, journalists, and industry experts. There is a saying that there is nothing new in science and research. While much of this book is based on my own experience as a United States Marine and education in psychology, it could not have been accomplished without the tireless work of others who have put in the time and effort to examine the psychological phenomena discussed in this text, as well as those that have left behind the memoirs and works of their own personal experiences.

Chapter 1 Resources

Bricout, Shirley. 'Cain: Figure of Rebellion and Resilience in D. H. Lawrence's tragic age', *Études Lawrenciennes* 52 (2021). <https://journals.openedition.org/lawrence/2316#quotation>.

Hyden, Marc. *Romulus: The Legend of Rome's Founding Father*. Pen And Sword History, 2020. <https://books.google.com/books?hl=en&lr=&id=w9MoEAAAQBAJ&oi=fnd&pg=PP1&dq=roman+mythology+romulus+and+remus+conflict&ots=TeGQ9drpZ4&sig=L8vbzZdXclREINec03GKcJhTmLc#v=onepage&q&f=false>.

Keegan, John. *A History of Warfare*. New York: Alfred A. Knopf, 1993.

Kissel, Marc and Nam C. Kim. The emergence of human warfare: Current perspectives. *American Journal of Physical Anthropology* 168.S67 (2018): 141-163. <https://onlinelibrary.wiley.com/doi/full/10.1002/ajpa.23751>.

LeShan, Lawerence. *The Psychology of War: Comprehending Its Mystique and Madness*. Expanded . New York: Helios Press, 2002.

LeShan, Lawrence. *Why We Love War*. 22 October 2007. <https://www.utne.com/community/why-we-love-war/>.

Muniandy, Tamil Arasi, Rajantheran Muniandy and Fonny Dameaty Hutagalung. Timeless gem: how Mahabharata can be the game changer in inspiring respect for elders. *Muslim Journal of Social Sciences and Humanities* 8.3 (2024): 135-152. <https://mjsshonline.com/index.php/journal/article/view/566>.

Strange, Deryn and Melanie K. T. Takarangi. 'Memory Distortion for Traumatic Events: The Role of Mental Imagery.' *Frontiers in Psychiatry* 6 (2015): 27. <https://pmc.ncbi.nlm.nih.gov/articles/PMC4337233/>.

Wilson, Edward O. *The Social Conquest of Earth*. W. W. Norton and Co., 2012. <http://sackett.net/e-o-wilson_the-social-conquest-of-earth.pdf>.

Chapter 2 Resources

Beckner, M., Main, L., Tait, J. L., Martin, B. J., Conkright, W. R., & Nindle, B. C. (2022). Circulating biomarkers associated with performance and resilience during military operational stress. *European Journal of Sport Science, 22*(1), 72-86. doi:https://doi.org/10.1080/17461391.2021.1962983

Chu, B., Marwaha, K., Sanvictores, T., Awosika, A. O., & Ayers, D. (2024, May 7). *Physiology, Stress Reaction*. Retrieved from StatPearls: https://www.ncbi.nlm.nih.gov/books/NBK541120/

Grossman, D., & Christensen, L. W. (2022). *On Combat: The Psychology and Physiology of Deadly Conflict in War and Peace*. Open Road Media. Retrieved from https://books.google.com/books?hl=en&lr=&id=ucx8EAAAQBAJ&oi=fnd&pg=PT8&dq=long-term+health+impacts+combat+stress+physiology&ots=tmg19YJ7q3&sig=VtwRoPhgPcScY-OFvv1ho_gib7M#v=onepage&q=long-term%20health%20impacts%20combat%20stress%20physiology&f=false

Hinds, J. A., & Sanchez, E. R. (2022). The Role of the Hypothalamus–Pituitary–Adrenal (HPA) Axis in Test-Induced Anxiety: Assessments, Physiological Responses, and Molecular Details. *Stresses, 2*(1), 146-155. doi:https://doi.org/10.3390/stresses2010011

Somvanshi, P., Mellon, S. H., FLory, J. D., Abu-Amara, D., Wolkowitz, O. M., Yehuda, R.. . . . Doyle, F. J. (2019). Mechanistic inferences on metabolic dysfunction in posttraumatic stress disorder from an integrated model and multiomic analysis: role of glucocorticoid receptor sensitivity. *American Journal of Physiology Endocrinology and Metabolism, 317*(5), E879-E898. doi:https://doi.org/10.1152/ajpendo.00065.2019

Sullivan-Kwantes, W., Cramer, M., Bouak, F., & Goodman, L. (2022). *Environmental stress in military settings*. Springer International Publishing. Retrieved from https://link.springer.com/content/pdf/10.1007/978-3-030-02866-4_107-1.pdf

Chapter 3 Resources

Acciarini, Chiara, Federica Brunetta and Paolo Boccardelli. 'Cognitive biases and decision-making strategies in times of change: a systematic literature review.' *Management Decision* 59.3 (2021). <https://www.emerald.com/insight/content/doi/10.1108.md-07-2019-1006/full/html>.

Bar-Tal, Daniel and Boaz Hameiri. 'Interventions to change well-anchored attitudes in the context of intergroup conflict.' *Social and Personality Psychology Compass* 14.7 (2020). <https://psycnet.apa.org/doi/10.1111/spc3.12534>.

Brown, Rupert and Samuel Pehrson. *Group Processes: Dynamics within and Between Groups*. 3rd. Hoboken: John Wiley & Sons, 2019. books.google.com.

Grimmell, Jan. *The Invisible Wounded Warriors in a Nation at Peace : An interview study on the lives of Swedish veterans of foreign conflicts and their experiences with PTSD, moral injuries, and military identities*. Casemate Group, 2022. www.torrossa.com. <ttps://www.torrossa.com/gs/resourceProxy?an=5622408&publisher=FZT653.>.

Malešević, Siniša. 'The moral fog of war and historical sociology.' *European Journal of Social Theory* 26.4 (2023): 490-501. <https://journals.sagepub.com/doi/full/10.1177/13684310231165218>.

Tarabay, Jennifer and Dennis Golm. 'The transmission of Intergenerational trauma and protective factors in survivors of the lebanese civil war and their adult offspring.' *International Journal of Intercultural Relations* 99 (2024): 101952. <https://www.sciencedirect.com/science/article/pii/S014717672400021X>.

Tryhorn, Dillon, et al. 'Modeling fog of war effects in AFSIM.' *The Journal of Defense Modeling and Simulation: Applications, Methodology, Technology* 20.2 (2021): 131-146. <https://journals.sagepub.com/doi/abs/10.1177/15485129211041963>.

Veronese, Guido, et al. 'Risk and Protective Factors Among Palestinian Women Living in a Context of Prolonged Armed Conflict and Political Oppression.' *Journal of Interpersonal Violence* 36.19-20 (2019): 9299-9327. <https://journals.sagepub.com/doi/abs/10.1177/0886260519865960>.

Chapter 4 Resources

Crane, Monique F., et al. 'Strengthening resilience in military officer cadets: A group-randomized controlled trial of coping and emotion regulatory self-reflection training.' *Journal of Consulting and Clinical Psychology* 87.2 (2019): 125-140.
<https://psycnet.apa.org/record/2018-58769-001>.

Delgado-Moreno, Rosa, et al. 'Effect of Experience and Psychophysiological Modification by Combat Stress in Soldier's Memory.' 43.150 (2019): 150.
<https://link.springer.com/article/10.1007/s10916-019-1261-1>.

Dexter, John C. 'Human resources challenges of military to civilian employment transitions.' *Career Development International* 25.5 (2020): 481-500.
<https://www.emerald.com/insight/content/doi/10.1108/cdi-02-2019-0032/full/html>.

Kendrick Jr., David. 'Why veterans feel addicted to combat.' *Psychology Today* (2022).
<https://www.psychologytoday.com/us/blog/the-veterans-corner/202201/why-veterans-feel-addicted-combat#:~:text=leave%20the%20battlefield.-,Being%20exposed%20to%20the%20adrenaline%20and%20the%20fame%20associated%20with,were%20in%20as%20a%20soldier.>.

Sledge, E.B. *With the Old Breed at Peleliu and Okinawa*. New York: Presidio Press, 2007.

Williamson, Victoria, et al. 'Moral injury: the effect on mental health and implications for treatment.' *The Lancet Psychiatry* 8.6 (2021): 453-455. <https://www.thelancet.com/journals/lanpsy/article/PIIS2215-0366(21)00113-9/fulltext>.

Chapter 5 Resources

Buheji , M., & Mushimiyimana , E. (2023). Raising Gaza survival capacity as per violence experienced lesson from best war survival stories in recent history. *International Journal of Advanced Research in Social Sciences and Humanities, 7*(1), 22-48. Doi:https://doi.org/10.17605/OSF.IO/YCBHU

Martin, P. A. (2020). Commander–community ties after civil war. *Journal of Peace Research, 58*(4), 778-793. Doi:https://doi.org/10.1177/0022343320929744

McCormick, W. H., Currier, J. M., Isaak, S. L., Sims, B. M., Slagel, B. A., Carroll, T. D.. . . . Albright, D. L. (2019). Military Culture and Post-Military Transitioning Among Veterans: A Qualitative Analysis. *Journal of Veterans Studies, 4*(2), 288-298. Doi:10.21061/jvs.v4i2.121

Senecal, G., McDonald, M. C., LaFleur, R., & Coey, C. (2019). Examining the effect of combat excitement & diminished civilian solidarity on life satisfaction for American veterans. *New Ideas in Psychology, 52*, 12-17. Doi:https://doi.org/10.1016/j.newideapsych.2018.09.001

Chapter 6 Resources

Alexander, Franz. 'Psychology and the interpretation of historical events.' *The Cultural Approach to History*. Ed. Caroline F. Ware. New York Chichester, West Sussex: Columbia University Press, 2019. 48-58.

Delgado-Moreno, Rosa, et al. 'Effect of Experience and Psychophysiological Modification by Combat Stress in Soldier's Memory.' 43.150 (2019): 150. <https://link.springer.com/article/10.1007/s10916-019-1261-1>.

Fitz, Hope K. *Ahimsa: A Way of Life; A Path to Peace*. University of Massachusetts Dartmouth Center for Indic Studies, 2007. <https://www.umassd.edu/media/umassdartmouth/center-for-indic-studies/gandhibooklet_2006.pdf>.

Fleischman, Paul. *The Buddha Taught Nonviolence, Not Pacifism*. 2002. <https://www.buddhistinquiry.org/article/the-buddha-taught-nonviolence-not-pacifism/>.

Randazzo, Elisa. 'The local, the 'Indigenous' and the limits of rethinking peacebuilding.' *Journal of Intervention and Statebuilding* 15.2 (2021): 141-160. <https://www.tandfonline.com/doi/abs/10.1080/17502977.2021.1882755>.

Tarabay, Jennifer and Dennis Golm. 'The transmission of Intergenerational trauma and protective factors in survivors of the lebanese civil war and their adult offspring.' *International Journal of Intercultural Relations* 99 (2024): 101952. <https://www.sciencedirect.com/science/article/pii/S014717672400021X>.

Veronese, Guido, et al. 'Risk and Protective Factors Among Palestinian Women Living in a Context of Prolonged Armed Conflict and Political Oppression.'*Journal of Interpersonal Violence* 36.19-20 (2019): 9299-9327. <https://journals.sagepub.com/doi/abs/10.1177/0886260519865960>.

Chapter 7 Resources

Bregman, Rutger. *Humankind: A Hopeful History*. Bloomsbury Publishing, 2020. books.google.com. <https://books.google.com/books?hl=en&lr=&id=UoYIEAAAQBAJ&oi=fnd&pg=PP1&dq=human+nature+lessons+learned+from+war&ots=H8QiCEjVHT&sig=Zbz-LeShan, Lawerence. *The Psychology of War: Comprehending Its Mystique and Madness*. Expanded . New York: Helios Press, 2002.

Heinrich, J., Bauer, M., Cassar, A., Chytilová, J., & Purzycki, B. G. (2019). War increases religiosity. *Nature Human Behaviour,*, *3*, 129-135. doi:https://doi.org/10.1038/s41562-018-0512-3

Puljek-Shank, Amela. *Healing and resiliency in a post-war setting*. 2012. <https://emu.edu/now/peacebuilder/2012/05/healing-and-resiliency-in-a-post-war-setting/>.

Chapter 8 Resources

Ainspan, Nathan D., Karen A. Orvis and Lynn M. Kelley. 'In navigating the complex transition from military service to civilian life, the concept of "embrace the suck" proves instrumental for veterans. This mindset, cultivated in challenging environments, allows individuals to transform adversity into opportuni.' *Workforce Readiness and the Future of Work*. Ed. Fred Oswald, Tara S. Behrend and Lori Foster. 1st. New York, 2019. 92-109. <https://www.taylorfrancis.com/chapters/edit/10.4324/9781351210485-6/military-source-civilian-workforce-development-nathan-ainspan-karin-orvis-lynne-kelley>.

Jackson, R. (2023, June 2). *Rebuilding Your Life and Finding Purpose After Addiction*. Retrieved from Recovery Centers of America: https://recoverycentersofamerica.com/blogs/rebuilding-your-life-and-finding-purpose-after-addiction/

McCormick, Wesley H., et al. 'Military Culture and Post-Military Transitioning Among Veterans: A Qualitative Analysis.' *Journal of Veterans Studies* 4.2 (2019): 288-298. <https://journal-veterans-studies.org/articles/10.21061/jvs.v4i2.121?utm_source=TrendMD&utm_medium=cpc&utm_campaign=Journal_of_Veterans_Studies_TrendMD_0>.

Weathers, Corie. *Military Culture Shift: The impact of war, money and generational perspective on moral, retention, and leadership*. Eva Resa Publishing, 2023. <https://books.google.com/books?hl=en&lr=&id=YQfaEAAAQBAJ&oi=fnd&pg=PA18&dq=embrace+the+suck+transformative+mindset+military+civilian+innovation&ots=vID9CnGy4l&sig=Z0Gw7G-rdK2kwBynT9rFb42a17w#v=onepage&q&f=false>.

Chapter 9 Resources

Coker, C. (2007). *The Warrior Ethos: Military Culture and the War on Terror* (1ST ed.). London: Routledge. doi:https://doi.org/10.4324/9780203089064

Gerhardt, T. F., Carlson, M., Moore, K. A., & Young, M. S. (2023). Veterans Treatment Courts: An Exploratory Analysis of the Effect of Veteran Mentors. *Criminal Justice Policy Review, 34*(6), 581-603. doi:https://doi.org/10.1177/08874034231202763

Pressfield, S. (2011). *The Warrior Ethos*. (S. Coyne, Ed.) Black Irish Entertainment LLC. Retrieved from https://books.google.com/books/about/The_Warrior_Ethos.html?id=OSXjAAAAQBA

About the Author

Nicholas Long is a combat veteran and father who also serves as a subject matter expert in instructional design and training development for government services, a professional coach, and is pursuing his PhD in Industrial-Organizational Psychology. His extensive career experience in military service, academia, government training development, and business finance leadership provides him with valuable insights that connect with both warriors and civilians. Adaptability and resilience have defined his life, and he remains committed to assisting others in discovering purpose and clarity during difficult times.

Using his personal battlefield experiences and background in psychology and mentoring, Nick investigates the mental and physical effects of combat. His work engages the relentless human spirit alongside practical methods for taking war lessons and using them to create meaningful leadership and impactful lives. As an industrial-organizational psychology PhD candidate, he utilizes the latest research methods to advance his knowledge of resilience and teamwork, as well as the power of purpose, across personal and professional contexts.

He passionately dedicates himself to fatherhood, where he gains daily inspiration from the values he works to teach his family. Nick demonstrates his dedication to legacy and warrior ethos through his combined influence as a parent and mentor. In all that he does, Nick brings the same intense focus and dedication to his current missions, guiding veterans through reintegration and helping professionals overcome the challenges that characterized his military service.

In his seminal work, Nick explores how war shapes identity and personal transformation. His writing provides readers with an empathetic and thoughtful examination of soldiers' transitions back to civilian life. His writing pays tribute to the warrior community's sacrifices and struggles while supplying practical guidance for those looking to rediscover their life's purpose. His coaching practice, along with academic work and storytelling, helps others discover inner strength during tough times and guides them toward meaningful life directions.